ISBN 978-1-330-09503-4
PIBN 10024450

1 MONTH OF
FREE
READING

at

www.ForgottenBooks.com

By purchasing this book you are eligible for one month membership to ForgottenBooks.com, giving you unlimited access to our entire collection of over 700,000 titles via our web site and mobile apps.

To claim your free month visit:
www.forgottenbooks.com/free24450

English
Français
Deutsche
Italiano
Español
Português

www.forgottenbooks.com

Mythology Photography **Fiction**
Fishing Christianity **Art** Cooking
Essays Buddhism Freemasonry
Medicine **Biology** Music **Ancient**
Egypt Evolution Carpentry Physics
Dance Geology **Mathematics** Fitness
Shakespeare **Folklore** Yoga Marketing
Confidence Immortality Biographies
Poetry **Psychology** Witchcraft
Electronics Chemistry History **Law**
Accounting **Philosophy** Anthropology
Alchemy Drama Quantum Mechanics
Atheism Sexual Health **Ancient History**
Entrepreneurship Languages Sport
Paleontology Needlework Islam
Metaphysics Investment Archaeology
Parenting Statistics Criminology
Motivational

PREFACE

It was the intention of the late Agnes Clerke to write the preface to this ' Children's Book of Stars.' Miss Clerke took a warm and sympathetic interest in the authoress and her work, but her lamented death occurred before this kindly intention could be fulfilled.

I cannot pretend to write adequately as her substitute, but I could not resist the appeal made to me by the author, in the name and for the sake of her dear friend and mine, to write a few words of introduction.

I am in no way responsible either for the plan or for any portion of this work, but I can commend it as a book, written in a simple and pleasant style, calculated to awaken the interest of intelligent children, and to enable parents otherwise ignorant of astronomy to answer many of those puzzling questions which such children often put.

DAVID GILL.

AUTHOR'S NOTE

THIS little work is the outcome of many suggestions on the part of friends who were anxious to teach their small children something of the marvels of the heavens, but found it exceedingly difficult to get hold of a book wherein the intense fascination of the subject was not lost in conventional phraseology —a book in which the stupendous facts were stated in language simple enough to be read aloud to a child without paraphrase.

Whatever merit there may be in the present work is due entirely to my friend Agnes Clerke, the well-known writer on astronomy ; the faults are all my own. She gave me the impetus to begin by her warm encouragement, and she helped me to continue by hearing every chapter read as it was written, and by discussing its successor and making suggestions for it. Thus she heard the whole book in MS. A week after the last chapter had been read to her I started on a journey lasting many months, and while I was in the Far East the news reached me of her death, by which the world is the poorer. For her sake, as he has stated, her friend Sir David Gill, K.C.B., kindly undertook to supply the missing preface.

G. E. MITTON.

CONTENTS

CONTENTS

ILLUSTRATIONS

PRINTED IN COLOUR

ILLUSTRATIONS

IN BLACK AND WHITE

THE CHILDREN'S BOOK
OF STARS

CHAPTER I

THE EARTH

IT is a curious fact that when we are used to things we often do not notice them, and things which we do every day cease to attract our attention. We find an instance of this in the curious change that comes over objects the further they are removed from us. They grow smaller and smaller, so that at a distance a grown-up person looks no larger than a doll; and a short stick planted in the ground only a few feet away appears as long as a much longer one at ten times the distance. This process is going on all round us every minute : houses, trees, buildings, animals, all seem larger or smaller in proportion to their distance from us. Sometimes I have seen a row of raindrops hanging on a bar by the window. When the sun catches one of them, it shines so brilliantly that it is as dazzling as a star; but my sense tells me it is a raindrop, and not a star at all. It is only because it is so near it seems as bright and important as a mighty star very, very far away.

We are so much accustomed to this fact that we get into a habit of judging the distance of things by their size. If we see two lights shining on a dark night, and one is much larger than the other, we think that the bright one must be nearer to us; yet it need not necessarily be so, for the two lights might possibly be at the same distance from us, and one be large and the other small. There is no way in which we can tell the truth by just looking at them. Now, if we go out on any fine moonlight night and look up at the sky, we shall see one object there apparently much larger than any other, and that is the moon, so the question that occurs to us at once is, Is the moon really very much larger than any of the stars, or does it only seem so because it is very much nearer to us? As a matter of fact, the moon is one of the smallest objects in view, only, as it is our nearest neighbour, it appears very conspicuous. Having learned this, we shall probably look about to see what else there is to attract attention, and we may notice one star shining very brilliantly, almost like a little lamp, rather low down in the sky, in that part of it where the sun has lately set. It is so beautifully bright that it makes all the others look insignificant in comparison, yet it is not really large compared with the others, only, as it comes nearer to us than anything else in the sky except the moon, it looks larger than it has any right to do in comparison with the others.

After this we might jump to the conclusion that all the bright large stars are really small and near to us, and all the faintly shining ones large and far away. But that would not be true at all, for some bright ones are very far away and some faint ones comparatively near, so that all we can do is to learn about them from the people who have studied them and found out about them, and then we shall know of our own knowledge which of them seem bright only because they are nearer than the others, and which are really very, very brilliant, and so still shine brightly, though set in space at an almost infinite distance from us.

The sun, as we all know, appears to cross the sky every day; he gets up in the east and drops down in the west, and the moon does the same, only the moon is unlike the sun in this, that it changes its shape continually. We see a crescent moon growing every night larger and larger, until it becomes full and fat and round, and then it grows thinner and thinner, until it dies away ; and after a little while it begins again, and goes through all the same changes once more. I will tell you why this is so further on, when we have a chapter all about the moon.

If you watch the stars quietly for at least five minutes, you will see that they too are moving steadily on in the same way as the sun and moon. Watch one bright star coming out from behind a chimney-pot, and after about five minutes you will see that it has changed its place. Yet this is not true of all, for if we watch carefully we shall find that some, fairly high up in the sky, do not appear to move at all. The few which are moving so slowly that they seem to us to stand still are at a part of the sky close to the Pole Star, so called because it is always above the North Pole of the earth. I will explain to you how to find it in the sky for yourselves later on, but now you can ask anyone to point it out. Watch it. It appears to be fixed in one place, while the other stars are swinging round it in circles. In fact, it is as if we on the earth were inside a great hollow globe or ball, which continually turned round, with the Pole Star near the top of the globe ; and you know that if you put your finger on the spot at the top of a spinning globe or ball, you can hold it there while all the rest of the ball runs round. Now, if you had to explain things to yourself, you would naturally think : 'Here is the great solid earth standing still, and the sun and moon go round it; the stars are all turning round it too, just as if they were fixed on to the inside of a hollow globe; we on the earth are in the middle looking up at them ; and this great globe is slowly wheeling round us night by night.'

In the childhood of the world men believed that this was really true—that the earth was the centre of the universe, that the sun and moon and all the hosts of heaven were there solely to light and benefit us ; but as the world grew wiser the wonders of creation were fathomed little by little. Some men devoted their whole lives to watching the heavens, and the real state of things was gradually revealed to them. The first great discovery was that of the daily movement of the earth, its rotation on its own axis, which makes it appear as if all these shining things went round it. It is indeed a very difficult matter to judge which of two objects is moving unless we can compare them both with something outside. You must have noticed this when you are sitting in a train at a station, and there is another train on the other side of yours. For if one of the trains moves gently, either yours or the other, you cannot tell which one it is unless you look at the station platform ; and if your position remains the same in regard to that, you know that your train is still standing, while the other one beside it has begun to move. And I am quite sure that there is no one of us who has not, at one time or another, stood on a bridge and watched the water running away underneath until we felt quite dizzy, and it seemed as if the water were standing still and the bridge, with ourselves on it, was flying swiftly away backwards. It is only when we turn to the banks and find them standing still, that we realize the bridge is not moving, and that it is the running water that makes it seem to do so. These everyday instances show us how difficult it is to judge whether we are moving or an outside object unless we have something else to compare with it. And the marvellous truth is that, instead of the sun and moon and stars rolling round the earth, it is the earth that is spinning round day by day, while the sun and the stars are comparatively still; and, though the moon does move, yet when we see her get up in the east and go down in the west that is due to our own movement and not to hers.

The earth turns completely round once in a day and night. If you take an orange and stick a knitting-needle through it, and hold it so that the needle is not quite straight up but a little slanting, and then twirl it round, you will get quite a good idea of the earth, though of course there is no great pole like a gigantic needle stuck through it, that is only to make it easy for you to hold it by. In spinning the orange you are turning it as the earth turns day by day, or, as astronomers express it, as it rotates on its axis.

There is a story of a cruel Eastern King who told a prisoner that he must die if he did not answer three questions correctly, and the questions were very difficult; this is one of them:

'How long would it take a man to go round the earth if he never stopped to eat or drink on the way?'

And the prisoner answered promptly: 'If he rose with the sun and kept pace with it all day, and never stopped for a moment to eat or drink, he would take just twenty-four hours, Your Royal

Highness.' For in those days it was supposed that the sun went round the earth.

Everyone is so remarkably clever nowadays that I am sure there will be someone clever enough to object that, if what I have said is true, there would be a great draught, for the air would be rushing past us. But, as a matter of fact, the air goes with us too. If you are inside a railway carriage with the windows shut you do not feel the rush of air, because the air in the carriage travels with you; and it is the same thing on the earth. The air which surrounds the earth clings to it and goes round with it, so there is no continuous breeze from this cause.

But the spinning round on its own axis is not the earth's only movement, for all the time it is also moving on round the sun, and once in a whole year it completes its journey and comes back to the place from whence it started. Thus the turning round like a top or rotating on its axis makes the day and night, and the going in a great ring or revolving round the sun makes the years.

Our time is divided into other sections besides days and years. We have, for instance, weeks and months. The weeks have nothing to do with the earth's movements; they are only made by man to break up the months; but the months are really decided by something over which we have no control. They are due to the moon, and, as I have said already, the moon must have a chapter to herself, so we won't say any more about the months here.

If any friend of ours goes to India or New Zealand or Australia, we look upon him as a great traveller; yet every baby who has lived one year on the earth has travelled millions of miles without the slightest effort. Every day of our lives we are all flung through space without knowing it or thinking of it. It is as if we were all shut up in a comfortable travelling car, and were provided with so many books and pictures and companions that we never cared to look out of the windows, so that hour by hour as we were carried along over miles of space we never gave them a thought. Even the most wonderful car ever made by man rumbles and creaks and shakes, so that we cannot help knowing it is moving; but this beautiful travelling carriage of ours called the earth makes never a creak or groan as she spins in her age-long journey. It is always astonishing to me that so few people care to look out of the window as we fly along; most of them are far too much absorbed in their little petty daily concerns ever to lift their eyes from them. It is true that sometimes the blinds are down, for the sky is thickly covered with clouds, and we cannot see anything even if we want to. It is true also that we cannot see much of the scenery in the daytime, for the sun shining on the air makes a veil of blue glory, which hides the stars; but on clear nights we can see on every side numbers of stars quite as interesting and beautiful as any landscape; and yet millions of people never look up, never give a thought to the wonderful scenery through which their car is rushing.

By reason of the onward rush of the earth in space we are carried over a distance of at least eighteen miles every second. Think of it: as we draw a breath we are eighteen miles away in space from the point we were at before, and this goes on unceasingly day and night. These astonishing facts make us feel how small and feeble we are, but we can take comfort in the thought that though our bodies are insignificant, the brain of man, which has discovered these startling facts, must in itself be regarded as one of the most marvellous of all the mysteries amid which we live.

Well, we have arrived at some idea of our earth's

position; we know that the earth is turning round day by day, and progressing round the sun year by year, and that all around lie the sentinel stars, scattered on a background of infinite space. If you take an older boy or girl and let him or her stand in the middle to represent the sun, then a smaller one would be the earth, and the smallest of all the moon; only in truth we could never get anyone large enough to represent the sun fairly, for the biggest giant that ever lived would be much too small in proportion. The one representing the sun must stand in the middle, and turn slowly round and round. Then let the earth-child turn too, and all the time she is spinning like a top she must be also hastening on in a big ring round the sun; but she must not go too fast, for the little moon-child must keep on running round her all the time. And the moon-child must keep her face turned always to the earth, so that the earth never sees her back. That is an odd thing, isn't it? We have never seen the other side of the moon, which goes round us, always presenting the same face to us.

The earth is not the only world going round the sun; she has many brothers and a sister; some are nearer to the sun than she is, and some are further away, but all circle round the great central light-giver in rings lying one outside the other. These worlds are called planets, and the earth is one of them, and one of the smaller ones, too, nothing so great and important as we might have imagined.

CHAPTER II

IF you are holding something in your hand and you let it go, what happens? It falls to the ground, of course. Now, why should it do so? You will say: 'How could it do anything else?' But that is only because you are hampered by custom. Try to shake yourself free, and think, Why should it go down instead of up or any other way? The first man who was clever enough to find some sort of an answer to this question was the great philosopher Sir Isaac Newton, though he was not quite the first to be puzzled by it. After years of study he discovered that every thing attracts every other thing in proportion to their masses (which is what you know as weight) and their distance from each other. In more scientific language, we should say every *body* instead of every *thing*, for the word body does not only mean a living body, but every lump or mass of matter in the universe. The earth is a body in this

13

sense, and so is the table or anything else you could name. Now as the earth is immeasurably heavier than anything that is on it, it pulls everything toward itself with such force that the little pulls of other things upon each other are not noticed. The earth draws us all toward it. It is holding us down to it every minute of the day. If we want to move we have to exert another force in order to overcome this attraction of the earth, so we exert our own muscles and lift first one foot and then the other away from the earth, and the effort we make in doing this tires us. All the while you are walking or running you are exercising force to lift your feet away from the ground. The pull of the earth is called gravitation. Just remember that, while we go on to something else which is almost as astonishing.

We know that nothing here on earth continues to move for ever ; everything has to be kept going, Anything left to itself has a tendency to stop. Why is this? This is because here in the world there is something that fights against the moving thing and tries to stop it, whether it be sent along the ground or thrown up in the air. You know what friction is, of course. If you rub your hands along any rough substance you will quickly feel it, but on a smooth substance you feel it less. That is why if you send a stone spinning along a carpet or a rough road it stops comparatively soon, whereas if you use the same amount of force and send it along a sheet of ice it goes on moving much longer. This kind of resistance, which we call friction, is one of the causes which is at work to bring things to a standstill; and another cause is the resistance of the air, which is friction in another form. It may be a perfectly still day, yet if you are bicycling you are breaking through the air all the time, just as you would be through water in swimming, only the resistance of the air is less than that of water. As the friction or the resistance of the air, or both combined, gradually lessens the pace of the stone you sent off with such force, the gravitation of the earth becomes apparent. When the stone first started the force you gave to it was enough to overcome the gravitation force, but as the stone moves more slowly the earth-pull asserts itself, and the stone drops down to the ground and lies still upon the surface. Now, if there were no friction, and therefore no resistance, there would be no reason why anything once set moving should not go on moving for ever. The force you give to any object you throw is enough to overcome gravitation ; and it is only when the first force has been diminished by the resistance of the air that the earth asserts its authority and pulls the moving object towards it.   If it were possible to get outside the air and out of reach of the pull of the earth, we might fling a ball off into space, and it would go on in a straight line until something pulled it to itself by the force of gravity.

Gravitation affects everything connected with the earth ; even our air is held to the earth by gravitation. It grows thinner and thinner as we get further away from the earth. At the top of a high mountain the air is so thin that men have difficulty in breathing, and at a certain height they could not breathe at all. As they cannot breathe in very fine air, it is impossible for them to tell by personal experiment exactly where the air ends ; but they have tried to find out in other ways, and though different men have come to different conclusions on the subject, it is safe to say that at about two hundred miles above the earth there is nothing that could be called air. Thus we can now picture our spinning earth clothed in a garment of air that clings closely about her, and grows thinner and thinner until it melts away altogether, for there is no air in space.

Now in the beginning God made the world, and set it off by a first impulse. We know nothing about the details, though further on you shall hear what is generally supposed to have taken place; we only know that, at some remote age, this world, probably very different from what it is now, together with the other planets, was sent spinning off into space on its age-long journey. These planets were not sent off at random, but must have had some particular connection with each other and with the sun, for they all belong to one system or family, and act and react on each other. Now, if they had been at rest and not in movement, they would have fallen right into the sun, drawn by the force of gravitation; then they would have been burned up, and there would have been an end of them. But the first force had imparted to them the impulse to go on in a straight line, so when the sun pulled the result was a movement between the two: the planets did not continue to move in a straight line, neither did they fall on to the sun, but they went on a course between the two—that is, a circle—for the sun never let them get right away from him, but compelled them to move in circles round him. There is a very common instance of this kind of thing which we can

2

see, or perhaps feel, every day. If you try to sit
still on a bicycle you tumble off, because the earth
pulls you down to itself; but if, by using the force
of your own muscles, you give the bicycle a forward
movement this resists the earth-pull, and the result
is the bicycle runs along the ground. It does not
get right away from the earth, not even two or three
feet above ground; it is held to the earth, but still
it goes forward and does not fall over, for the movement is made up of the earth-pull, which holds it
to the ground, and the forward movement, which
propels it along. Then again, as another instance,
if you tie a ball to a string and whirl it round you,
so long as you keep on whirling it will not fall to
the ground, but the moment you stop down it
drops, for there is nothing to fight against the
pull of gravitation. Thus we can picture the
earth and all the planets as if they were swinging
round the sun, held by invisible strings. It is the
combination of two forces that keeps them in their
places—the first force and the sun's pull. It is very
wonderful to think of. Here we are swinging in
space on a ball that seems only large to us because
we are so much smaller ourselves; there is nothing
above or below it but space, yet it travels on
day by day and year by year, held by invisible

forces that the brain of man has discovered and measured.

Of course, every planet gives a pull at every other planet too, but these pulls are so small compared with that of the sun that we need not at present notice them.   Then we come to another point.   We said that every body pulled every other body in proportion to their weights and their distance.   Now, gravity acts much more strongly when things are near together than when they are far away from each other; so that if a smaller body is near to another somewhat larger than itself, it is pulled by it much more strongly than by a very much larger one at a considerably greater distance.   We have an instance of this in the case of the earth and moon: as the earth responds to the pull of the sun, so the moon responds to the pull of the earth.   The moon is so comparatively near to the earth that the earth-pull forces her to keep on going round and round, instead of leaving her free to circle round the sun by herself; and yet if you think of it the moon does go round the sun too.   Recall that game we had when the sun was in the middle, and the two smaller girls, representing the earth and moon, went round it.   The moon-child turned round the earth-child, but all the while the earth-child was going round the sun, so that in a year's time the moon had been all round the sun too, only not in a straight line. The moon is something like a dog who keeps on dancing round and round you when you go for a walk. He does go for the walk too, but he does much more than that in the same time. Thus we have further completed our idea of our world. We see it now hanging in space, with no visible support, held in its place by two mighty forces; spinning on year after year, attended by its satellite the moon, while we run, and walk, and cry, and laugh, and play about on its surface—little atoms who, except for the brain that God has given them, would never even have known that they are continually moving on through endless space.

CHAPTER III

THE SHINING MOON

'ONCE upon a time,' long, long ago, the earth was not a compact, round, hard body such as she is now, but much larger and softer, and as she rotated a fragment broke off from her; it did not go right away from her, but still went on circling round with the motion it had inherited from her. As the ages passed on both the earth and this fragment, which had been very hot, cooled down, and in cooling became smaller, so that the distance between them was greater than it had been before they shrank. And there were other causes also that tended to thrust the two further from each other. Yet, compared with the other heavenly bodies, they are still near, and by looking up into the sky at night you can generally see this mighty fragment, which is a quarter the diameter of the earth—that is to say, a quarter the width of the earth measured from side to side through the middle. It is—as, of course, you have guessed

—the moon. The moon is the nearest body to us in all space, and so vast is the distance that separates us from the stars that we speak as if she were not very far off, yet compared with the size of the earth the space lying between us and her is very great. If you went right round the world at the thickest part—that is to say, in the region of the Equator—and when you arrived at your starting-point went off once again, and so on until you had been round ten times, you would only then have travelled about as far as from the earth to the moon!

The earth is not the only planet which has a moon, or as it is called, a satellite, in attendance. Some of the larger planets have several, but there is not one to compare with our moon. Which would you prefer if you had the choice, three or four small moons, some of them not much larger than a very big bright star, or an interesting large body like our own moon? I know which I should say.

'You say that the moon broke off from the earth, so perhaps there may be some people living on her,' I hear someone exclaim.

If there is one thing we have found out certainly about the moon, it is that no life, as we know it, could exist there, for there is neither air nor water. Whether she ever had any air or water, and if so, why they disappeared, are questions we cannot answer. We only know that now she is a dead world. Bright and beautiful as she is, shedding on us a pale, pure light, in vivid contrast with the fiery yellow rays of the sun, yet she is dead and lifeless and still. We can examine her surface with the telescope, and see it all very plainly. Even with a large opera-glass those markings which, to the naked eye, seem to be like a queer distorted face are changed, and show up as the shadows of great mountains. We can only see one side of the moon, because as I have said, she keeps always the same face turned to the earth; but as she sways slightly in her orbit, we catch a glimpse of sometimes a little more on one side and sometimes a little more on the other, and so we can judge that the unseen part is very much the same as that turned toward us.

At first it is difficult to realize what it means to have no air. Besides supporting life in every breath that is drawn by living creatures, the air does numerous other kind offices for us—for instance, it carries sound. Supposing the most terrific volcano exploded in an airless world, it could not be heard. The air serves as a screen by day to keep off the burning heat of the sun's rays, and as a blanket by night to keep in the heat and not let it escape too quickly. If there were no air there could be no water, for all water would evaporate and vanish at once. Imagine the world deprived of air; then the sun's rays would fall with such fierceness that even the strongest tropical sun we know would be as nothing in comparison with it, and every green thing would shrivel up and die; this scorching sun would shine out of a black sky in which the stars would all be visible in the daytime, not hidden by the soft blue veil of air, as they are now. At night the instant the sun disappeared below the horizon black darkness would set in, for our lingering twilight is due to the reflection of the sun in the upper layers of air, and a bitterness of deathly cold would fall upon the earth—cold fiercer than that of the Arctic regions—and everything would be frozen solid. It would need but a short time to reduce the earth to the condition of the moon, where there is nothing to shrivel up, nothing to freeze. Her surface is made up of barren, arid rocks, and her scenery consists of icy black shadows and scorching white plains.

The black shadows define the mountains, and tremendous mountains they are. Most of them

have craters. A crater is like a cup, and generally has a little peak in the middle of it. This is the summit of a volcano, and when the volcano has burst up and vomited out floods of lava and débris, this has fallen down in a ring a little distance away from it, leaving a clear space next to the peak, so that, as the mountain ceases vomiting and the lava cools down, the ring hardens and forms a circular ridge. The craters on the moon are immense, not only in proportion to her size, but immense even according to our ideas on the earth. One of the largest craters in our own world is in Japan, and this measures seven miles across, while in the moon craters of fifty, sixty, and even a hundred miles are by no means uncommon, though there are also hundreds and thousands of smaller ones. We can see the surface of the moon very plainly with the magnificent telescopes that have now been made, and with the best of these anything the size of a large town would be plainly visible. Needless to say, no town ever has been or ever will be seen upon the moon !

All these mountains and craters show that at one time the moon must have been convulsed with terrific disturbances, far worse than anything that we have any knowledge of on our earth ; but this must have been ages ago, while the moon still probably had an atmosphere of its own. Now it has long been quiet. Nothing changes there; even the forces that are always at work on the earth—namely, damp and mould and water—altering the surface and breaking up the rocks, do not act there, where there is no moisture of any sort. So far as we can see, the purpose of the moon is to be the servant of the earth, to give her light by night and to raise the tides. Beautiful light it is, soft and mysterious—light that children do not often have a chance of seeing, for they are generally in bed before the moon rises when she is at the full.

We know that the moon has no heat of her own—she parted with all that long ago; she cannot give us glowing light from brilliant flames, as the sun does; she shines only by the reflection of the sun on her surface, and this is the reason why she appears to change her shape so constantly. She does not really change; the whole round moon is always there, only part of it is in shadow. Sometimes you can see the dark part as well as the bright. When there is a crescent moon it looks as if it were encircling the rest; some people call it, 'seeing the old moon in the new moon's arms.' I don't know if you would guess why it is we can see the dark part then, or how it is lighted up. It is by reason of our own shining, for we give light to the moon, as she does to us. The sun's rays strike on the earth, and are reflected on to the moon, so that the moon is lighted by earthshine as we are lighted by moonshine, and it is these reflected earth-rays that light up the dark part of the moon and enable us to see it. What a journey these rays have had! They travel from the sun to the earth, and the earth to the moon, and then back to the earth again! From the moon the earth must appear a much bigger and more glorious spectacle than she does to us—four times wider across and probably brighter—for the sun's light strikes often on our clouds, which shine more brilliantly than her surface.

Once again we must use an illustration to explain the subject. Set a lamp in the middle of a dark room, and let that be the sun, then take a small ball to represent the earth and a smaller one for the moon. Place the moon-ball between the lamp and the earth-ball. You will see that the side turned to the earth-ball is dark, but if you move the moon to one side of the earth, then from the earth half of it appears light and half dark; if you put it right away from the lamp, on the outer side of the earth, it is all gloriously lit up, unless it happens to be exactly behind the earth, when the earth's shadow will darken it.   This is the full explanation of all the changes of the moon.

Does it ever fall within the earth's shadow?   Yes, it does; for as it passes round the earth it is not always at the same level, but sometimes a little higher and sometimes a little lower, and when it

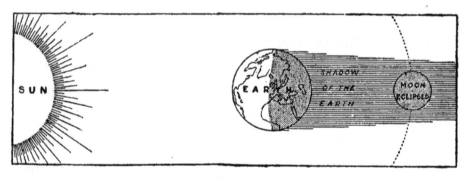

AN ECLIPSE OF THE MOON.

chances to pass exactly behind it enters the shadow and disappears.   That is what we call an eclipse of the moon.   It is nothing more than the earth's shadow thrown on to the moon, and as the shadow is round that is one of the proofs that the earth is round too.   But there is another kind of eclipse—the eclipse of the sun; and this is caused by the moon herself.   For when she is nearest to the sun, at new moon—that is to say, when her dark side is toward us, and she happens to get exactly between us and the sun—she shuts out the face of the sun from us; for though she is tiny compared with him, she is so much nearer to us that she appears almost the same size, and can blot him right out. Thus the eclipses of both sun and moon are not difficult to understand: that of the moon can only happen at full moon, when she is furthest from the sun, and it is caused by the earth's shadow falling upon the

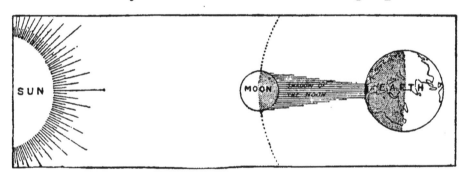

AN ECLIPSE OF THE SUN.

moon; and that of the sun at new moon, when she is nearest to him, and it is caused by the solid body of the moon coming between us and the sun.

Besides giving us light by night, the moon serves other important purposes, and the most important of all is the raising of the tides. Without the rising of the sea twice in every day and night our coasts would become foul and unwholesome, for all the dead fish and rotting stuff lying on the beach would

poison the air.   The sea tides scour our coasts day
by day with never-ceasing energy, and they send a
great breath of freshness up our large rivers to
delight many people far inland.   The moon does
most of this work, though she is a little helped by
the sun.   The reason of this is that the moon is
so near to the earth that, though her pull is a

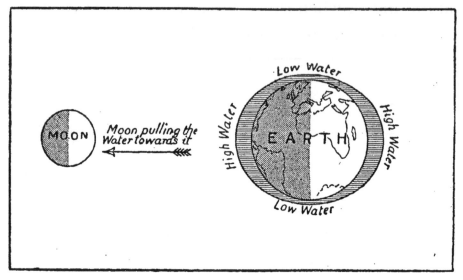

THE MOON RAISING THE TIDES.

comparatively small one, it is very strongly felt.
She cannot displace the actual surface to any great
extent, as it is so solid ; but when it comes to the
water she can and does displace that, so that the
water rises up in answer to her pull, and as the earth
turns round the raised-up water lags behind, reaching backward toward the moon, and is drawn up

on the beach, and makes high tide. But it is stopped there, and meantime, by reason of the earth's movement, the moon is left far behind, and pulls the water to itself further on, when the first high tide relapses and falls down again. At length the moon gets round to quite the opposite side of the earth to that where she began, and there she makes a high tide too; but as she draws the water to herself she draws also the solid earth beneath the water to her in some degree, and so pulls it away from the place where the first high tide occurred, leaving the water there deeper than before, and so causing a secondary high tide.

The sun has some influence on the tides too, and when moon and sun are in the same line, as at full and new moon, then the tides are highest, and are called spring tides; but when they pull in different directions, as when it is half-moon, then the tides are lowest and are called neap tides.

CHAPTER IV

THE EARTH'S BROTHERS AND SISTER

THE earth is not the only world that, poised in space, swings around the sun. It is one of a family called the Solar System, which means the system controlled and governed by the sun. When we look up at the glorious sky, star-studded night by night, it might seem to us that the stars move only by reason of the earth's rotation; but when men first began to study the heavens attentively—and this is so long ago that the record of it is not to be found—they noticed that, while every shining object in the sky was apparently moving round us, there were a few which also had another movement, a proper motion of their own, like the moon. These curious stars, which appeared to wander about among the other stars, they called planets, or wanderers. And the reason, which was presently discovered, of our being able to see these movements was that these planets are very much nearer to us than any of the real

32

stars, and in fact form part of our own solar system, while the stars are at immeasurable distances away.   Of all the objects in the heavens the planets are the most intensely interesting to us ; for though removed from us by millions of miles, the far-reaching telescope brings some of them within such range that we can see their surfaces and discover their movements in a way quite impossible with the stars.   And here, if anywhere, might we expect to find traces of other living beings like ourselves ; for, after all the earth is but a planet, not a very large nor a very small one, and in no very striking position compared with the other planets ; and thus, arguing by what seems common-sense, we say, If this one planet has living beings on its surface, may not the other planets prove to be homes for living beings also ?   Counting our own earth, there are eight of these worlds in our solar system, and also a number of tiny planets, called asteroids ; these likewise go round the sun, but are very much smaller than any of the first eight, and stand in a class by themselves, so that when the planets are mentioned it is generally the eight large well-known planets which are referred to.

If we go back for a moment to the illustration of the large lamp representing our sun, we shall now be able to fill in the picture with much more detail. The orbits of the planets, as their paths round the sun are called, lie like great circles one outside another at various distances, and do not touch or cut each other. Where do you suppose our own place to be? Will it be the nearest to the sun or the furthest away from him? As a matter of fact, it is neither, we come third in order from the sun, for two smaller planets, one very small and the other nearly as large as the earth, circle round and round the sun in orbits lying inside ours. Now if we want to place objects around our lamp-sun which will represent these planets in size, and to put them in places corresponding to their real positions, we should find no room large enough to give us the space we ought to have. We must take the lamp out into a great open field, where we shall not be limited by walls. Then the smallest planet, named Mercury, which lies nearest of all to the sun, would have to be represented by a pea comparatively close to the sun; Venus, the next, would be a greengage plum, and would be about twice as far away; then would come the earth, a slightly larger plum, about half as far again as Venus. After this there would be a lesser planet, called Mars, like a marble. These are the first

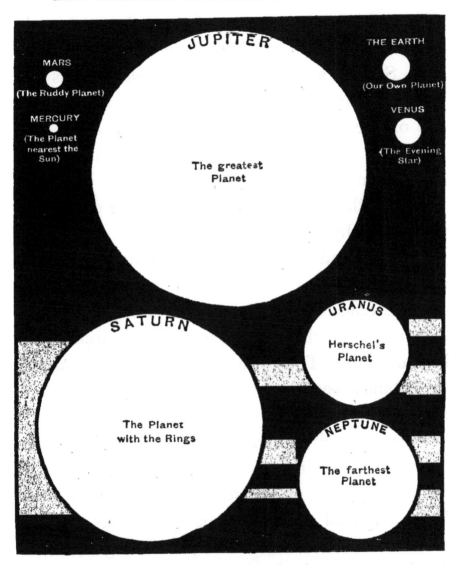

COMPARATIVE SIZES OF THE PLANETS.

four, all comparatively small; beyond them there
is a vast gap, in which we find the asteroids, and
after this we come to four larger planets, mighty
indeed as regards ourselves, for if our earth were

3—2

a greengage plum, the first of these, Jupiter, would have to be the size of a football at least, and the next, Saturn, a smaller football, while Uranus and Neptune, the two furthest out, would be about the size of the toy balloons children play with. The outermost one, Neptune, would be thirty times as far from the sun as we are.

This is the solar system, and in it the only thing that shines by its own light is the sun; all the rest, the planets and their moons, shine only because the rays of light from the sun strike on their surfaces and are reflected off again. Our earth shines like that, and from the nearer planets must appear as a brilliant star. The little solar system is separated by distances beyond the realm of thought from the rest of the universe. Vast as are the intervals between ourselves and our planetary neighbours, they are as nothing to the space that separates us from the nearest of the steady shining fixed stars. Why, removed as far from us as the stars, the sun himself would have sunk to a point of light; and as for the planets, the largest of them, Jupiter, could not possibly be seen. Thus, when we look at those stars across the great gulf of space, we know that though we see them they cannot see us, and that to them our sun must seem only a star; con-

sequently we argue that perhaps these stars themselves are suns with families of planets attached to
them; and though there are reasons for thinking
that this is not the case with all, it may be with
some.   Now if, after learning this, we look again
at the sky, we do so with very different eyes, for
we realize that some of these shining bodies are
like ourselves in many things, and are shining only
with a light borrowed from the sun, while others
are mighty glowing suns themselves, shining by
their own light, some greater and brighter, some
less than our sun.   The next thing to do is to learn
which are stars and which are planets.

Of the planets you will soon learn to pick out
one or two, and will recognize them even if they
do change their places—for instance, Venus is at
times very conspicuous, shining as an evening star
in the west soon after the sun goes down, or as
a morning star before he gets up, though you are
not so likely to see her then; anyway, she is never
found very far from the sun.   Jupiter is the only
other planet that compares with her in brilliancy,
and he shines most beautifully.   He is, of course,
much further away from us than Venus, but so
much larger that he rivals her in brightness.
Saturn can be quite easily seen as a conspicuous

object, too, if you know where to look for him, and
Mars is sometimes very bright with a reddish glow.
The others you would not be able to distinguish.

It is to our earth's family of these eight large
planets going steadily round the same sun that we
must give our attention first, before going on to
the distant stars.   Many of the planets are accom-
panied by satellites or moons, which circle round
them.   We may say that the sun is our parent—
father, mother, what you will—and that the planets
are the family of children, and that the moons are
their children.   Our earth, you see, has only one
child, but that a very fine one, of which she may
well be proud.

When I say that the planets go round the sun in
circles I am only speaking generally ; as a matter
of fact, the orbits of the planets are not perfect
circles, though some are more circular than others.
Instead of this they are as a circle might look if it
were pressed in from two sides, and this is called an
ellipse.   The path of our own earth round the sun
is one of the most nearly circular of them all, and
yet even in her orbit she is a good deal nearer to
the sun at one time than another.   Would you be
surprised to hear that she is nearer in our winter
and further away in our summer ?   Yet that is the

case.   And for the first moment it seems absurd ;
for what then makes the summer hotter than the
winter ?   That is due to an altogether different
cause ; it depends on the position of the earth's
axis.   If that axis were quite straight up and down
in reference to the earth's path round the sun we
should have equal days and nights all the year
round, but it is not ; it leans over a little, so that at
one time the North Pole points towards the sun and
at another time away from it, while the South Pole
is pointing first away from it and then toward it in
exactly the reverse way.   When the North Pole
points to the sun we in the Northern Hemisphere
have our summer.   To understand this you must look
at the picture, which will make it much clearer than
any words of mine can do.   The dark part is the
night, and the light part the day.   When we are
having summer any particular spot on the Northern
Hemisphere has quite a long way to travel in the
light, and only a very short bit in the dark, and the
further north you go the longer the day and shorter
the night, until right up near the North Pole, within
the Arctic Circle, it is daylight all the time.   You
have, perhaps, heard of the 'midnight sun' that
people go to see in the North, and what the expres-
sion means is that at what should be midnight the

sun is still there.   He seems just to circle round the horizon, never very far above, but never dipping below it.

When the sun is high overhead, his rays strike down with much more force than when he is low.   It is, for instance, hotter at mid-day than in the evening.   Now, when the North Pole is bowed toward the sun, the sun appears to us to be higher in the sky. In the British Isles he never climbs quite to the zenith, as we call the point straight above our heads ; he always keeps on the southern side of that, so that our shadows are thrown northward at mid-day, but yet he gets nearer to it than he does in winter. Look at the picture of the earth as it is in winter. Then we have long nights and short days, and the sun never appears to climb very high, because we are turned away from him.   During the short days we do not receive a great deal of heat, and during the long night the heat we have received has time to evaporate to a great extent.   These two reasons— the greater or less height of the sun in the sky and the length of the days—are quite enough to account for the difference between our summer and winter. There is one rather interesting point to remember, and that is that in the Northern Hemisphere, whether it is winter or summer, the sun is south at mid-day,

so that you can always find the north then, for your shadow will point northwards.

New Zealand and Australia and other countries placed in the Southern Hemisphere, as we are in the Northern, have their summer while we have winter, and winter while we have summer, and their summer is warmer than ours, because it comes when the earth in its journey is three million miles nearer to the sun than in our summer.

All this seems to refer to the earth alone, and this chapter should be about the planets ; but, after all, what applies to one planet applies to another in some degree, and we can turn to the others with much more interest now to see if their axes are bowed toward the sun as ours is. It is believed that in the case of Mercury, in regard to its path round the sun, the axis is straight up and down ; if it is the changes of the seasons must depend on the nearness of Mercury to the sun and nothing else, and as he is a great deal nearer at one time than another, this might make a very considerable difference. Some of the planets are like the earth in regard to the position of their axes, but the two outermost ones, Uranus and Neptune, are very peculiar, for one pole is turned right toward the sun and the other right away from it, so that in one hemisphere there is continuous day all the summer, in the other there is continuous night, and then the process is reversed. But these little peculiarities we shall have to note more particularly in the account of the planets separately.

There is a curious fact in regard to the distances of the planets from the sun. Each one, after the first, is, very roughly, about double the distance from the sun of the one inside it. This holds good for all the first four, then there is a great gap where we might expect to find another planet, after which follow the four large planets. Now, this gap puzzled astronomers greatly; for though there seemed to be no reason why the planets should be at regular distances one outside the other, yet there the fact was, and that the series should be broken by a missing planet was annoying. So very careful search was made, and a thrill of excitement went all through the scientific world when it was known that a tiny planet had been discovered in the right place. But this was not the end of it, for within a few years three or four more tiny planets were observed not far from the first one, and, as years rolled on, one after another was discovered until now the number amounts to over six hundred and others are perpetually being added to the list! Here was a new feature in the solar system, a band of tiny planets not one of which was to be compared in size with the least of those already known. The largest may be about as large as Europe, and others perhaps about the size of Wales, while there may be many that have only a few square miles of surface altogether, and are too small for us to see. To account for this strange discovery many theories were advanced.

One was that there had been a planet—it might be about the size of Mars—which had burst up in a great explosion, and that these were the pieces—a very interesting and exciting idea, but one which proved to be impossible. The explanation now generally accepted is a little complicated, and to understand it we must go back for a bit.

When we were talking of the earth and the moon we realized that once long ago the moon must have been a part of the earth, at a time when the earth was much larger and softer than she now is; to put it in the correct way, we should say when she was less dense. There is no need to explain the word 'dense,' for in its ordinary sense we use it every day, but in an astronomical sense it does not mean exactly the same thing. Everything is made up of minute particles or atoms, and when these atoms are not very close together the body they compose is loose in texture, while if they are closer together the body is firmer. For instance, air is less dense than water, and water than earth, and earth than steel. You see at once by this that the more density a thing has the heavier it is; for as a body is attracted to another body by every atom or particle in it, so if it has more particles it will be more strongly attracted. Thus on the earth the denser things are really heavier. But 'weight' is only a word we use in connection with the earth; it means the earth's pulling power toward any particular thing at the surface, and if we were right out in space away from the earth, the pulling power of the earth would be less, and so the weight would be less; and as it would be impossible always to state just how far away a thing was from the earth, astronomers talk about density, which means the number of particles a body contains in proportion to other bodies. Thus the planet Jupiter is very much larger than the earth, but his density is less. That does not mean to say that if Jupiter were in one scale and the earth in the other he would weigh less, because he is so very much bigger he would outweigh the earth still ; his total *mass* would be greater than that of the earth, but it means that a piece of Jupiter the same size as a piece of the earth would weigh less under the same conditions.

Now, before there were any planets at all or any sun, in the place of our solar system was a vast gaseous cloud called a nebula, which slowly rotated, and this rotation was the first impulse or force which God gave it. It was not at all dense, and as it rotated a part broke off, and inheriting the first impulse, went on rotating too. The impulse would have sent it off in a straight line, but the pull of gravity from the nebula held it in place, and it circled round ; then the nebula, as it rotated, contracted a little, and occupied less space and grew denser, and presently a second piece was thrown off, to become in time another planet. The same process was repeated with Saturn, and then with the huge Jupiter. The nebula was always rotating and always contracting. And as it behaved, so did the planets in their turn ; they spun round and cooled and contracted, and the moons were flung off from them, just as they—the planets —had been flung off from the parent nebula.

Now, after the original nebula had parted with the mighty mass of Jupiter, it never again made an effort so great, and for a long time the fragments that were detached were so small as hardly to be worth calling planets; they were the asteroids, little lumps and fragments that the nebula left behind. But as it still contracted in time there came Mars; and having recovered a little, the nebula with more energy got rid of the earth, and next Venus, and lastly little Mercury, the smallest of the eight planets. Then it contracted further, and perhaps you can guess what the remainder of it is—the sun; and by spinning in a plastic state the sun, like the earth, has become a globe, round and comparatively smooth; and its density is now too great to allow of its losing any more fragments, so, as far as we can see, the solar system is complete.

This theory of the origin of the planets is called the nebula theory. We cannot prove it, but there are so many facts that can only be explained by it, we have strong reason for believing that something of the kind must have happened. When we come to speak of the starry heavens we shall see that there are many masses of glowing gas which are nebulæ of the same sort, and which form an object-lesson in our own history.

We have spoken rather lightly of the nebula rotating and throwing off planets; but we must not think of all this as having happened in a short time. It is almost as impossible for the human mind to conceive the ages required for such slow changes as to grasp the great gulfs of space that separate us from the stars. We can only do it by comparison. You know what a second is, and how the seconds race past without ceasing day and night. It makes one giddy to picture the seconds there are in a year ; yet if each one of those seconds was a year in itself, what then ? That seems a stupendous time, but it is nothing compared with the time needed to form a nebula into a planetary system. If we had five thousand of such years, with every second in them a year, we should then only have counted one billion real years, and billions must have passed since the sun was a gaseous nebula filling the outermost bounds of our system !

CHAPTER V

FOUR SMALL WORLDS

WHAT must the sun appear to Mercury, who is so much nearer to him than we are? To understand that we should have to imagine our sun increased to eight or nine times his apparent size, and pouring out far greater heat and light than anything that we have here, even in the tropics. It was at first supposed that Mercury must have an extra thick covering of clouds to protect him from this tremendous glare; but recent observations tend to prove that, far from this, he is singularly free from cloud. As this is so, no life as we know it could possibly exist on Mercury.

His year—the time he takes to go round the sun and come back to the same place again—is eighty-eight days, or about one-quarter of ours. As his orbit is much more like an ellipse than a circle, it follows that he is much nearer to the sun at one time than at another—in fact, when he is nearest, the size of the sun must seem three and a half

48

times greater than when he is furthest away from it ! Even at the best Mercury is very difficult to observe, and what we can learn about him is not much ; but, as we have heard, his axis is supposed to be upright. If so his seasons cannot depend on the bend toward or away from the sun, but must be influenced solely by the changes in his distance from the sun, which are much greater than in our own case. There is some reason to believe, too, that Mercury's day and year are the same length. This means that as the planet circles round the sun he turns once. If this is so the sun will shine on one half of the planet, producing an accumulated heat terrific to think of; while the other side is plunged in blackness. The side which faces the sun must be heated to a pitch inconceivable to us during the nearer half of the orbit—a pitch at which every substance must be at boiling-point, and which no life as we know it could possibly endure. Seen from our point of view, Mercury goes through all the phases of the moon, as he shines by the reflected light of the sun ; but this point we shall consider more particularly in regard to Venus, as Venus is nearer to us and easier to study. For a long time astronomers had a fancy that there might be another planet even nearer to the sun

4

than Mercury, perhaps hidden from us by the great glare of the sun. They even named this imaginary planet Vulcan, and some thought they had seen it, but it is tolerably certain that Vulcan existed only in imagination. Mercury is the nearest planet to the sun, and also the smallest, of course excepting the asteroids. It is about three thousand miles in diameter, and as our moon is two thousand miles, it is not so much bigger than that. So far as we are concerned, it is improbable we shall ever know very much more about this little planet.

But next we come to Venus, our beautiful bright neighbour, who approaches nearer to us than any other heavenly body except the moon. Alas! when she is nearest, she like Mercury, turns her dark side toward us, coming in between us and the sun, so that we cannot observe her at all.

Everyone must have noticed Venus, however carelessly they have looked at the sky; but it is likely that far more people have seen her as an evening than a morning star, for most people are in bed when the sun rises, and it is only before sunrise or after sunset we can see Venus well. She is at her best from our point of view when she seems to us to be furthest from the sun, for then we can study her best, and at these times she appears like a half or three-quarter moon, as we only see a part of the side from which the sunlight is reflected.   She shines like a little silver lamp,

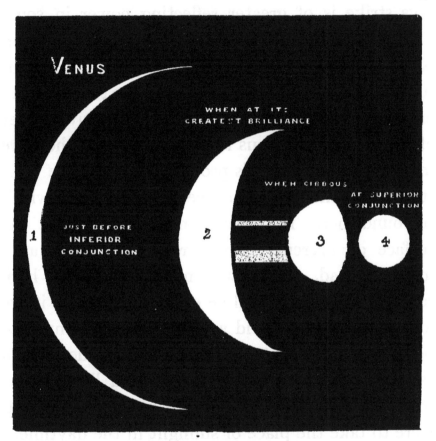

WHEN AT ITS GREATEST BRILLIANCE

WHEN GIBBOUS

AT SUPERIOR CONJUNCTION

VENUS

JUST BEFORE INFERIOR CONJUNCTION

1

2

3

4

DIFFERENT PHASES OF VENUS.

excelling every other planet, even Jupiter, the largest of all.   If we look at her even with the naked eye, we can see that she is elongated or drawn out, but her brilliance prevents us from seeing her shape exactly; to do this we must use a telescope.

It is a curious fact that some planets shine much more brightly than others, without regard to their size—that is to say, the surface on which the sun's rays strike is of greater reflecting power in some than in others. One of the brightest things in Nature that we can imagine is a bank of snow in sunlight ; it is so dazzling that we have to look away or wink hard at the sight ; and the reflective power of the surface of Venus is as dazzling as if she were made of snow. This is probably because the light strikes on the upper surface of the clouds which surround her. In great contrast to this is the surface of Mercury, which reflects as dully as a mass of lead. Our own moon has not a high reflecting power, as will be easily understood if we imagine what the world would be if condemned to perpetual moonlight only. It would, indeed, be a sad deprivation if the mournful cold light of the moon, welcome enough as a change from sunlight, were to take the place of sunlight in the daytime.

For a very long time astronomers could not discover what time Venus took in rotating on her own axis—that is to say, what the length of her day was. She is difficult to observe, and in order to find out the rotation it is necessary to note some fixed object on the surface which turns round with the planet and comes back to the same place again, so that the time it takes in its journey can be measured. But the surface of Venus is always changing, so that it is impossible to judge at all certainly. Opinions differ greatly, some astronomers holding that Venus's day is not much longer than an earthly day, while others believe that the planet's day is equal to her year, just as in the case of Mercury. Venus's year is 225 days, or about seven and a half of our months, and if, indeed, her day and year are the same length, very peculiar effects would follow. For instance, terrible heat would be absorbed by the side of the planet facing the sun in the perpetual summer; and the cold which would be felt in the dreary winter's night would far exceed our bitterest Arctic climate. We cannot but fancy that any beings who might live on a planet of this kind must be different altogether from ourselves. Then, there is another point: even here on earth very strong winds are caused by the heating of the tropics; the hot air, being lighter than the cold air, rises, and the colder air from the poles rushes in to supply its place. This causes wind, but the winds which would be raised on Venus by the rush of air from the icy side of the planet to the hot one would be tornadoes such as we could but faintly dream of. It is, of course, useless to speculate when we know so little, but in a subject so intensely interesting we cannot help guessing a little.

Venus is only slightly smaller than the earth, and her density is not very unlike ours; therefore the pull of gravity must be pretty much there what it is here—that is to say, things will weigh at her surface about the same as they do here. Her orbit is nearly a circle, so that her distance from the sun does not vary much, and the heat will not be much greater from this cause at one time of the year than another.

As her orbit is tilted up a little she does not pass between us and the sun at each revolution, but occasionally she does so, and this passing is called a transit. Many important facts have been learned by watching these transits. Mercury also has transits across the sun, but as he is so much smaller than Venus they are not of such great importance. It was by the close observation of Venus during her transits that the distance from the earth to the sun was first measured. Not until the year 2004 will another transit of Venus occur.

It is not difficult to imagine that the earth must appear a splendid spectacle from Venus, whence she is seen to great advantage. When nearest to us she must see us like a little moon, with markings as the continents and seas rotate, and these will

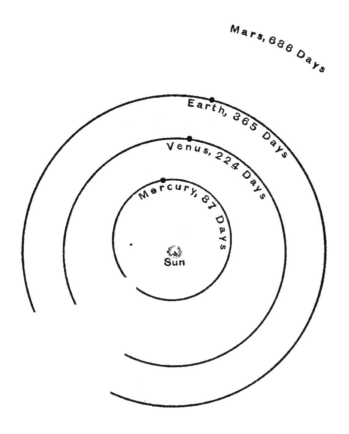

ORBITS OF MARS, THE EARTH, VENUS, AND MERCURY.

change as they are obscured by the clouds rolling over them. At the North and South Poles will be glittering ice-caps, growing larger and smaller as they turn toward or away from the sun.   A brilliant spectacle !

We might say with a sigh, 'If only we could see such a world !' Well, we can see a world—not indeed, so large as Venus, yet a world that comes almost as near to us as Venus does, and which, unlike her, is outside us in order from the sun, so that when it is nearest to us the full sunlight is on it.   This is Mars, our neighbour on the other side, and of all the fascinating objects in the sky Mars is the most fascinating, for there, if anywhere, should we be likely to discover beings like ourselves !

Mars takes rather more than half an hour longer to rotate than we do, and as he is so much smaller than the earth, this means that he moves round more slowly.   His axis is tilted at nearly the same angle as ours is.   Mars is much smaller than the earth, his diameter is about twice that of the moon, and his density is about three-quarters that of the earth, so that altogether, with his smaller size and less density, anything weighing a hundred pounds here would only weigh some forty pounds on Mars; and if, by some miraculous agency, you were suddenly transported there, you would find yourself so light that you could jump enormous distances with little effort, and skip and hop as if you were on springs.

Look at the map of Mars, in which the surface appears to be cut up into land and water, continents and oceans. The men who first observed Mars with accuracy saw that some parts were of a reddish colour and others greenish, and arguing from our own world, they called the greenish parts seas and the reddish land. For a long while no one doubted that we actually looked on a world like our own, more especially as there was supposed to be a covering of atmosphere. The so-called land and water are much more cut up and mixed together than ours, it is true. Here and there is a large sea, like that marked 'Mare Australe,' but otherwise the water and the land are strangely intermingled. The red colour of the part they named land puzzled astronomers a good deal, for our land seen at the same distance would not appear so red, and they came at last to the conclusion that vegetation on Mars must be red instead of green! But after a while another disturbing fact turned up to upset their theories, and that was that they saw canals, or what they called canals, on Mars. These were long, straight, dark markings, such as you see on the map. It is true that some people never saw these markings at all, and disbelieved in their existence; but others saw them clearly, and watched them change—first go fainter and then darker again. And quite recently a photograph has been obtained which shows them plainly, so they must have an existence, and cannot be only in the eye of the observer, as the most sceptical people were wont to suggest. But further than this, one astronomer announced that some of these lines appeared to be double, yet when he looked at them again they had grown single. It was like a conjuring trick. Great excitement was aroused by this, for if the canals were altered so greatly it really did look as if there were intelligent beings on Mars capable of working at them. In any case, if these are really canals, to make them would be a stupendous feat, and if they are artificial—that is, made by beings and not natural—they show a very high power of engineering. Imagine anyone on earth making a canal many miles wide and two thousand miles long! It is inconceivable, but that is the feat attributed to the Martians. The supposed doubling of the canals, as I say, caused a great deal of talk, and very few people could see that they were double at all. Even now the fact is doubted, yet there seems every reason to believe it is true. They do not all appear to be double, and those that do are always the same ones, while others undoubtedly remain single all the time. But the canals do not exhaust the wonders of Mars. At each pole there is an ice-cap resembling those found at our own poles, and this tells us pretty plainly something about the climate of Mars, and that there is water there.

This ice-cap melts when the pole which it surrounds is directed toward the sun, and sometimes in a hot summer it dwindles down almost to nothing, in a way that the ice-caps at the poles of the earth never do. A curious appearance has been noticed when it is melting: a dark shadow seems to grow underneath the edge of it and extends gradually, and as it extends the canals near it appear much darker and clearer than they did before, and then the canals further south undergo the same change. This looks as if the melting of the snow filled up the canals with water, and was a means of watering the planet by a system totally different from anything we know here, where our poles are surrounded by oceans, and the ice-caps do not in the least affect our water-supply. But, then, another strange fact had to be taken into consideration. These straight lines called canals ran out over the seas occasionally, and it was impossible to believe that if they were canals they could do that. Other things began to be discussed, such as the fact that the green parts of Mars did not always remain green. In what is the springtime of Mars they are so, but afterwards they become yellow, and still later in the season parts near the pole turn brown. Thus the idea that the greenish parts are seas had to be quite given up, though it appeared so attractive. The idea now generally believed is that the greenish parts are vegetation—trees and bushes and so on, and that the red parts are deserts of reddish sand, which require irrigation—that is to say, watering—before anything can be grown on them. The apparent doubling of the canals may be due to the green vegetation springing up along the banks. This might form two broad lines, while the canal itself would not be seen, and when the vegetation dies down, we should see only the trench of the canal, which would possibly appear faint and single. Therefore the arrangements on Mars appear to be a rich and a barren season on each hemisphere, the growth being caused by the melting of the polar ice-cap, which sends floods down even beyond the Equator. If we could

imagine the same thing on earth we should have
to think of pieces of land lying drear and dry
and dead in winter between straight canal-like
ditches of vast size. A little water might remain
in these ditches possibly, but not enough to water
the surrounding land. Then, as summer progressed,
we should hear, ' The floods are coming,' and each
deep, huge canal would be filled up with a tide
of water, penetrating further and further. The
water drawn up into the air would fall in dew
or rain. Vegetation would spring up, especially
near the canal banks, and instead of dreary wastes
rich growths would cover the land, gradually dying
down again in the winter. So far Mars seems in
some important respects very different from the
earth. He is also less favourably placed than we are,
for being so much further from the sun, he receives
very much less heat and light. His years are 687 of
our days, or one year and ten and a half months, and
his atmosphere is not so dense as ours. With this
greater distance from the sun and less air we might
suppose the temperature would be very cold indeed,
and that the surface would be frost-bound, not only
at the poles, but far down towards the Equator.
Instead of this being so, as we have seen, the polar
caps melt more than those on the earth. We can

only surmise there must be some compensation we do not know of that softens down the rigour of the seasons, and makes them milder than we should suppose possible.

Of course, the one absorbing question is, Are there people on Mars? To this it is at present impossible to reply. We can only say the planet seems in every way fitted to support life, even if it is a little different from our earth. It is most certainly a living world, not a dead one like the moon, and as our knowledge increases we may some day be able to answer the question which so thrills us.

Our opportunities for the observation of Mars vary very greatly, for as the earth's orbit lies inside that of Mars, we can best see him when we are between him and the sun. Of course, it must be remembered that the earth and the other planets are so infinitely small in regard to the space between them that there is no possibility of any one of them getting in such a position that it would throw a shadow on any other or eclipse it. The planets are like specks in space, and could not interfere with one another in this way. When Mars, therefore, is in a line with us and the sun we can see him best, but some of these times are better than others, for this reason—the earth's orbit is nearly a circle, and that of Mars more of an ellipse.

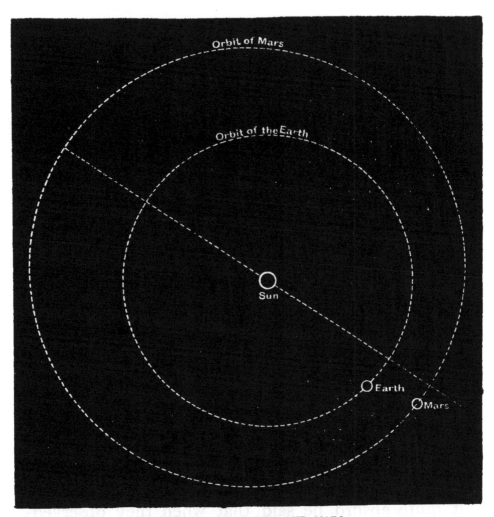

ORBITS OF THE EARTH AND MARS.

Look at the illustration and remember that Mars' year is not quite two of ours—that is to say, every time we swing round our orbit we catch him up in a different place, for he will have progressed less than half his orbit while we go right round ours.

Sometimes when we overtake him he may be at that part which is furthest away from us, or he may be at that part which is nearest to us, and if he is in the latter position we can see him best. Now at these, the most favourable times of all, he is still more than thirty-five millions of miles away—that is to say, one hundred and forty times as far as the moon, yet comparatively we can see him very well. He is coming nearer and nearer to us, and very soon will be nearer than he has been since 1892, or fifteen years ago. Then many telescopes will be directed on him, and much may be learned about him.

For a long time it was supposed that Mars had no moons, and when Dean Swift wrote 'Gulliver's Travels' he wanted to make the Laputans do something very clever, so he described their discovery of two moons attending Mars, and to make it quite absurd he said that when they observed these moons they found that one of them went round the planet in about ten hours. Now, as Mars takes more than twenty-four hours to rotate, this was considered ridiculous, for no moon known then took less time to go round its primary world than the primary world took to turn on its own axis. Our own moon, of course, takes thirty times as long—that is a month contains thirty days. Then one hundred and fifty years later this jest of Dean Swift's came true, for two moons were really discovered revolving round Mars, and one of them does actually take less time to complete its orbit than the planet does to rotate—namely, a little more than seven hours! So the absurdity in ' Gulliver's Travels' was a kind of prophecy!

These two moons are very small, the outer one perhaps five or six miles in diameter, and the inner one about seven; therefore from Mars the outer one, Deimos, cannot look much more than a brilliant star, and the inner one would be but a fifth part the apparent width of our own moon. So Mars is not very well off, after all. Still, there is great variety, for it must be odd to see the same moon appearing three times in the day, showing all the different phases as it goes from new to full, even though it is small!

Such wonderful discoveries have already been made that it is not too much to say that perhaps some day we may be able to establish some sort of communication with Mars, and if it be inhabited by any intelligent beings, we may be able to signal to them ; but it is almost impossible that any contrivance could bridge the gulf of airless space that separates us, and it is not likely that holiday trips to Mars will ever become fashionable !

CHAPTER VI

FOUR LARGE WORLDS

I HAVE told you about the four lesser worlds of which our earth is one, and you know that beyond Mars, the last of them, there lies a vast space, in which are found the asteroids, those strange small planets circling near to each other, like a swarm of bees. After this there comes Jupiter, who is so enormous, so superb in size compared with us, that he might well serve as the sun of a little system of his own. You remember that we represented him by a football, while the earth was only a greengage plum. But Jupiter himself is far less in comparison with the sun than we are in comparison with him. He differs from the planets we have heard about up to the present in that he seems to glow with some heat that he does not receive from the sun. The illumination which makes him appear as a star to us is, of course, merely reflected sunlight, and what we see is the external covering, his envelope of cloud.

There is every reason to believe that the great

bulk of Jupiter is still at a high temperature. We know that in the depths of the earth there is still plenty of heat, which every now and then makes its presence felt by bursting up through the vents we call volcanoes, the weak spots in the earth's crust; but our surface long ago cooled, for the outside of any body gets cool before the inside, as you may have found if ever you were trying to eat hot porridge, and circled round the edge of the plate with a spoon. A large body cools more slowly than a small one, and it is possible that Jupiter, being so much larger than we are, has taken longer to cool. One reason we have for thinking this is that he is so very light compared with his size— in other words, his density is so small that it is not possible he could be made of materials such as the earth is made of.

As I said, when we study him through telescopes we see just the exterior, the outer envelope of cloud, and as we should expect, this changes continually, and appears as a series of belts, owing to the rotation of the planet. Jupiter's rotation is very rapid; though he is so much greater than the earth, he takes less than half the time the earth does to turn round—that is to say, only ten hours. His days and nights of five hours each seem short to us, accustomed to measure things by our own estimates. But we must remember that everything is relative; that is to say, there is really no such thing as fast or slow; it is all by comparison. A spider runs fast compared with a snail, but either is terribly slow compared with an express train; and the speed of an express train itself is nothing to the velocity of light.

In the same way there is nothing absolutely great or small; it is all by comparison. We say how marvellous it is that a little insect has all the mechanism of life in its body when it is so tiny, but if we imagine that insect magnified by a powerful microscope until it appears quite large, the marvel ceases. Again, imagine a man walking on the surface of the earth as seen from a great distance through a telesope: he would seem less than an insect, and we might ask how could the mechanism of life be compressed into anything so small? Thus, when we say enormous or tiny we must always remember we are only speaking by the measurements of our own standards.

There is nothing very striking about Jupiter's orbit. He takes between eleven and twelve of our years to get round the sun, so you see, though his day is shorter, his year is longer than ours. And this is not only because his path is much larger, but because by the law of gravity the more distant a planet is from the sun the more slowly it travels, so that while the earth speeds over eighteen miles Jupiter has only done eight. Of course, we must be careful to remember the difference between rotation and revolution. Jupiter rotates much quicker than the earth—that is to say, he turns round more quickly—but he actually gets over the ground more slowly. The sun appears much smaller to him than it does to us, and he receives considerably less light and heat. There are various spots on his surface, and one remarkable feature is a dark mark, which is called the 'great red spot.' If as we suppose what we see of the planet is merely the cloudy upper atmosphere, we should not expect to find anything permanent there, for the markings would change from day to day, and this they do with this exception—that this spot, dark red in colour, has been seen for many years, turning as the planet turned. It was first noticed in 1878, and was supposed to be some great mountain or excrescence peeping up through the clouds. It grew stronger and darker for several years, and then seemed to fade, and was not so easily seen, and though still remaining it is now pale. But, most startling to say, it has shifted its position a little—that is, it takes a few seconds longer to get round the planet than it did at first. A few seconds, you will say, but that is nothing! It does not seem much, but it shows how marvellously accurate astronomers are. Discoveries of vast importance have been made from observing a few seconds' discrepancy in the time the heavenly bodies take in their journeys, and the fact that this spot takes a little longer in its rotation than it did at first shows that it cannot be attached to the body of the planet. It is impossible for it to be the summit of a mountain or anything of that sort. What can it be? No one has yet answered that question.

Jupiter has eight moons, the last of which was only discovered in 1908; it is very small and very far away from the planet. Four of the moons are quite large. They have the honour of having been the first heavenly bodies ever actually discovered, for the six large planets nearest the sun have been known so long that there is no record of their first discovery, and of course our own moon has always been known. Galileo, who invented the telescope, turned it on to the sky in 1610, when our King Charles I. was on the throne, and he saw these curious bodies which at first he could not believe to be moons. The four which he saw vary in size from two thousand one hundred miles in diameter to nearly three thousand six hundred. You remember our own moon is two thousand miles across, so even the smallest is larger than she. They go round at about the same level as the planet's Equator, and therefore they cross right in front of him, and go behind him once in every revolution. Since then the other three have been

JUPITER AND HIS PRINCIPAL MOONS.

discovered in the band of Jupiter's satellites—one a small moon closer to him than any of the first set, and two others further out. It was by observation of the first four, however, that very interesting results were obtained. Mathematicians calculated the time that these satellites ought to disappear behind Jupiter and reappear again, but they found that this did not happen exactly at the time predicted; sometimes the moons disappeared sooner than they should have done, and sometimes later. Then this was discovered to have some relation to the distance of our earth from Jupiter. When he was at the far side of his immense orbit he was much more distant from us than when he was on the nearer side—in fact, the difference may amount to more than three hundred millions of miles. And it occurred to some clever man that the irregularities in time we noticed in the eclipses of the satellites corresponded with the distance of Jupiter from us. The further he drew away from us, the later were the eclipses, and as he came nearer they grew earlier. By a brilliant inspiration, this was attributed to the time light took to travel from them to us, and this was the first time anyone had been able to measure the velocity or speed of light. For all practical purposes, on the earth's surface we hold light to be instantaneous, and well we may, for light could travel more than eight times round the world in one second. It makes one's brain reel to think of such a thing. Then think how far Jupiter must be away from us at the furthest, when you hear that sometimes these eclipses were delayed seventeen minutes—minutes, not seconds—because it took that time for light to cross the gulf to us!

Sound is very slow compared with light, and

that is why, if you watch a man hammering at a distance, the stroke he gives the nail does not coincide with the bang that reaches you, for light gets to you practically at once, and the sound comes after it.   No sound can travel without air, as we have heard, therefore no sound reaches us across space.   If the moon were to blow up into a million pieces we should see the amazing spectacle, but should hear nothing of it.   Light travels everywhere throughout the universe, and by the use of this universal carrier we have learnt all that we know about the stars and planets.   When the time that light takes to travel had been ascertained by means of Jupiter's satellites, a still more important problem could be solved—that was our own distance from the sun, which before had only been known approximately, and this was calculated to be ninety-two millions seven hundred thousand miles, though sometimes we are a little nearer and sometimes a little further away.

Jupiter is marvellous, but beyond him lies the most wonderful body in the whole solar system. We have found curiosities on our way out: we have studied the problem of the asteroids, of the little moon that goes round Mars in less time than Mars himself rotates ; we have considered the ' great

red spot' on Jupiter, which apparently moves independently of the planet ; but nothing have we
found as yet to compare with the rings of Saturn.
May you see this amazing sight through a telescope
one day !

Look at the picture of this wonderful system,
and think what it would be like if the earth
were surrounded with similar rings ! The first
question which occurs to all of us is what must
the sky look like from Saturn ? What must
it be to look up overhead and see several great
hoops or arches extending from one horizon to
another, reflecting light in different degrees of
intensity ? It would be as if we saw several
immense rainbows, far larger than any earthly
rainbow, and of pure light, not split into colours,
extending permanently across the sky, and now
and then broken by the black shadow of the
planet itself as it came between them and the sun.
However, we must begin at the beginning, and
find out about Saturn himself before we puzzle
ourselves over his rings. Saturn is not a very
great deal less than Jupiter, though, so small are
the other planets in comparison, that if Saturn and
all the rest were rolled together, they would not
make one mass so bulky as Jupiter ! Saturn is so

light—in other words, his density is so small—that he is actually lighter than water. He is the lightest, in comparison with his size, of any of the planets. Therefore he cannot be made largely of solid land, as our earth is, but must be to a great extent, composed of air and gaseous vapour, like his mighty neighbour. He approaches at times as near to Jupiter as Jupiter does to us, and on these occasions he must present a splendid spectacle to Jupiter. He takes no less than twenty-nine and a half of our years to complete his stately march around the sun, and his axis is a little more tilted than ours ; but, of course, at his great distance from the sun, this cannot have the same effect on the seasons that it does with us. Saturn turns fast on his axis, but not so fast as Jupiter, and in turning his face, or what we call his surface, presents much the same appearance to us that we might expect, for it changes very frequently and looks like cloud belts.

The marvellous feature about Saturn is, of course, the rings. There are three of these, lying one within the other, and separated by a fine line from each other. The middle one is much the broadest, probably about ten thousand miles in width, and the inner one, which is the darkest, was not discovered until some time after the others. As the planet swings in his orbit the rings naturally appear very different to us at different times. Sometimes we can only see them edgewise, and then even in the largest telescope they are only like a streak of light, and this shows that they cannot be more than fifty or sixty miles in thickness. The one which is nearest to Saturn's surface does not approach him within ten thousand miles. Saturn has no less than ten satellites, in addition to the rings, so that his midnight sky must present a magnificent spectacle. The rings, which do not shine by their own light but by reflected sunlight, are solid enough to throw a shadow on the body of the planet, and themselves receive his shadow. Sometimes for days together a large part of Saturn must suffer eclipse beneath the encircling rings, but at other times, at night, when the rings are clear of the planet's body, so that the light is not cut off from them, they must appear as radiant arches of glory spanning the sky.

The subject of these rings is so complicated by the variety of their changes that it is difficult for us even to think about it. It is one of the most marvellous of all the features of our planetary system. What are these rings? what are they made of? It has been positively proved that they cannot be made of continuous matter, either liquid or solid, for the force of gravity acting on them from the planet would tear them to pieces. What, then, can they be? It is now pretty generally believed that they are composed of multitudes of tiny bodies, each separate, and circling separately round the great planet, as the asteroids circle round the sun. As each one is detached from its neighbour and obeys its own impulses, there is none of the strain and wrench there would be were they all connected. According to the laws which govern planetary bodies, those which are nearest to the planet will travel more quickly than those which are further away. Of course, as we look at them from so great a distance, and as they are moving, they appear to us to be continuous. It is conjectured that the comparative darkness of the inside ring is caused by the fact that there are fewer of the bodies there to reflect the sunlight. Then, in addition to the rings, enough themselves to distinguish him from all other planets, there are the ten moons of richly-endowed Saturn to be considered. It is difficult to gather much about these moons, on account of our great distance from them. The largest is probably twice the diameter of our own moon. One of them seems to be much brighter—that is to say, of higher reflecting power—on one side than the other, and by distinguishing the sides and watching carefully, astronomers have come to the conclusion that it presents always the same face to Saturn in the same way as our own moon does to us; in fact, there is reason to think that all the moons of large planets do this.

All the moons lie outside the rings, and some at a very great distance from Saturn, so that they can only appear small as seen from him. Yet at the worst they must be brighter than ordinary stars, and add greatly to the variations in the sky scenery of this beautiful planet. In connection with Saturn's moons there is another of those astonishing facts that are continually cropping up to remind us that, however much we know, there is such a vast deal of which we are still ignorant. So far in dealing with all the planets and moons in the solar system we have made no remark on the way they rotate or revolve, because they all go in the same direction, and that direction is called counter-clockwise, which means that if you stand facing a clock and turn your hand slowly round the opposite direction to that in which the hands go, you will be turning it in the same way that the earth rotates on its axis and revolves in its orbit. It is, perhaps, just as well to give here a word of caution. Rotating of course means a planet's turning on its own axis, revolving means its course in its orbit round the sun. Mercury, Venus, Earth, Mars, Jupiter, and all their moons, as well as Saturn himself, rotate on their axes in this one direction—counter-clockwise—and revolve in the same direction as they rotate. Even the queer little moon of Mars, which runs round him quicker than he rotates, obeys this same rule. Nine of Saturn's moons follow this example, but one independent little one, which has been named Phœbe, and is far out from the planet, actually revolves in the opposite way. We cannot see how it rotates, but if, as we said just now, it turns the same face always to Saturn, then of course it rotates the wrong way too. A theory has been suggested to account for this curious fact, but it could not be made intelligible to anyone who has not studied rather high mathematics, so there we must just leave it, and put it in the cabinet of curiosities we have already collected on our way out to Saturn.

For ages past men have known and watched the planets lying within the orbit of Saturn, and they had made up their minds that this was the limit of our system. But in 1781 a great astronomer named Herschel was watching the heavens through a telescope when he noticed one strange object that he was certain was no star. The vast distance of the stars prevents their having any definite outline, or what is called a disc. The rays dart out from them in all directions and there is no 'edge' to them, but in the case of the planets it is possible to see a disc with a telescope, and this object which attracted Herschel's attention had certainly a disc. He did not imagine he had discovered a new planet, because at that time the asteroids had not been found, and no one thought that there could be any more planets. Yet Herschel knew that this was not a star, so he called it a comet! He was actually the first who discovered it, for he knew it was not a fixed star, but it was after his announcement of this fact that some one else, observing it carefully, found it to be a real planet with an orbit lying outside that of Saturn, then the furthest boundary of the solar system. Herschel suggested calling it Georgium Sidus, in honour of George III., then King; but luckily this ponderous name was not adopted, and as the other planets had been called after the Olympian deities, and Uranus was the father of Saturn, it was called Uranus. It was subsequently found that this new planet had already been observed by other astronomers and catalogued as a star no less than seventeen times, but until Herschel's clear sight had detected the difference between it and the fixed stars no one had paid any attention to it. Uranus is very far away from the sun, and can only sometimes be seen as a small star by people who know exactly where to look for him. In fact, his distance from the sun is nineteen times that of the earth.

Yet to show at all he must be of great size, and that size has actually been found out by the most delicate experiments. If we go back to our former comparison, we shall remember that if the earth were like a greengage plum, then Uranus would be in comparison about the size of one of those coloured balloons children play with; therefore he is much larger than the earth.

In this far distant orbit the huge planet takes eighty-four of our years to complete one of his own. A man on the earth will have grown from babyhood to boyhood, from boyhood to the prime of life, and lived longer than most men, while Uranus has only once circled in his path.

But in dealing with Uranus we come to another of those startling problems of which astronomy is full. So far we have dealt with planets which are more or less upright, which rotate with a rotation like that of a top. Now take a top and lay it on one side on the table, with one of its poles pointing toward the great lamp we used for the sun and the other pointing away. That is the way Uranus gets round his path, on his side! He rotates the wrong way round compared with the planets we have already spoken of, but he revolves the same way round the sun that all the others do. It seems wonderful that even so much can be found out about a body so far from us, but we know more: we have discovered that Uranus is made of lighter material than the earth; his density is less. How can that be known? Well, you remember every body attracts every other body in proportion to the atoms it contains. If, therefore, there were any bodies near to Uranus, it could be calculated by his influence on them what was his own mass, which, as you remember, is the word we use to express what would be weight were it at the earth's surface; and far away as Uranus is, the bodies from which such calculations may be made have been discovered, for he has no less than four satellites, or moons. Considering now the peculiar position of the planet, we might expect to find these moons revolving in a very different way from others, and this is indeed the case. They turn round the planet at about its Equator—that is to say, if you hold the top representing Uranus as was suggested just now, these moons would go above and below the planet in passing round it. Only we must remember there is really no such thing as above and below absolutely. We who are on one side of the world point up to the sky and down to the earth, while the people on the other side of the earth, say at New Zealand, also point up to the sky and down to the earth, but their pointings are directly the opposite of ours. So when we speak of moons going above and below that is only because, for the moment, we are representing Uranus as a top we hold in our hands, and so we speak of above and below as they are to us.

It was Herschel who discovered these satellites, as well as the planet, and for these great achievements he occupies one of the grandest places in the roll of names of which England is proud. But he did much more than this : his improvements in the construction of telescopes, and his devotion to astronomy in many other ways, would have caused him to be remembered without anything else.

Of Uranus's satellites one, the nearest, goes round in about two and a half days, and the one that is furthest away takes about thirteen and a half days, so both have a shorter period than our moon.

The discovery of Uranus filled the whole civilized world with wonder. The astronomers who had seen him, but missed finding out that he was a planet, must have felt bitterly mortified, and when he was discovered he was observed with the utmost accuracy and care. The calculations made to determine his path in the sky were the easier because he had been noted as a star in several catalogues previously, so that his position for some time past was known. Everybody who worked at astronomy began to observe him. From these facts mathematicians set to work, and, by abstruse calculations, worked out exactly the orbit in which he ought to move; then his movements were again watched, and behold he followed the path predicted for him; but there was a small difference here and there: he did not follow it exactly. Now, in the heavens there is a reason for everything, though we may not always be clever enough to find it out, and it was easily guessed that it was not by accident that Uranus did not precisely follow the path calculated for him. The planets all act and react on one another, as we know, according to their mass and their distance, and in the calculations the pull of Jupiter on Saturn and of Saturn on Uranus were known and allowed for. But Uranus was pulled by some unseen influence also.

A young Englishman named Adams, by some abstruse and difficult mathematical work far beyond the power of ordinary brains, found out not only the fact that there must be another planet nearly as large as Uranus in an orbit outside his, but actually predicted where such a planet might be seen if anyone would look for it. He gave his results to a professor of astronomy at Cambridge. Now, it seems an easy thing to say to anyone, 'Look out for a planet in such and such a part of the sky,' but in reality, when the telescope is turned to that part of the sky, stars are seen in such numbers that, without very careful comparison with a star chart, it is impossible to say which are fixed stars and which, if any, is an intruder. There happened to be no star chart of this kind for the particular part of the sky wanted, and thus a long time elapsed and the planet was not identified. Meantime a young Frenchman named Leverrier had also taken up the same investigation, and, without knowing anything of Adams' work, had come to the same conclusion. He sent his results to the Berlin Observatory, where a star chart such as was wanted was actually just being made. By the use of this the Berlin astronomers at once identified this new member of our system, and announced to the astonished world that another large planet, making eight altogether, had been discovered. Then the English astronomers remembered that they too held in their hands the means for making this wonderful discovery, but, by having allowed so much time to elapse, they had let the honour go to France. However, the names of Adams and Leverrier will always be coupled together as the discoverers of the new planet, which was called Neptune. The marvel is that by pure reasoning the mind of man could have achieved such results.

If the observation of Uranus is difficult, how much more that of Neptune, which is still further plunged in space ! Yet by patience a few facts have been gleaned about him. He is not very different in size from Uranus. He also is of very slight density. His year includes one hundred and sixty-five of ours, so that since his discovery in 1846 he has only had time to get round less than a third of his path. His axis is even more tilted over than that of Uranus, so that if we compare Uranus to a top held horizontally, Neptune will be like a top with one end pointing downwards.　He rotates in this extraordinary position, in the same manner as Uranus—namely, the other way over from all the other planets, but he revolves, as they all do, counter-clockwise.

Seen from Neptune the sun can only appear about as large as Venus appears to us at her best, and the light and heat received are but one nine-hundreth part of what he sends us.　Yet so brilliant is sunshine that even then the light that falls on Neptune must be very considerable, much more than that which we receive from Venus, for the sun itself glows, and from Venus the light is only reflected.　The sun, small as it must appear, will shine with the radiance of a glowing electric light.　To get some idea of the brilliance of sunlight, sit near a screen of leaves on some sunny day when the sun is high overhead, and note the intense radiance of even the tiny rays which shine through the small holes in the leaves.　The scintillating light is more glorious than any diamond, shooting out coloured rays in all directions.　A small sun the apparent size of Venus would, therefore, give enough light for practical purposes to such a world as Neptune, even though to us a world so illuminated would seem to be condemned to a perpetual twilight.

CHAPTER VII

THE SUN

So far we have referred to the sun just so much as was necessary to show the planets rotating round him, and to acknowledge him as the source of all our light and heat; but we have not examined in detail this marvellous furnace that nourishes all the life on our planet and burns on with un-diminished splendour from year to year, without thought or effort on our part. To sustain a fire on the earth much time and care and expense are necessary; fuel has to be constantly supplied, and men have to stoke the fire to keep it burning. Considering that the sun is not only vastly larger than all the fires on the earth put together, but also than the earth itself, the question very naturally occurs to us, Who supplies the fuel, and who does the stoking on the sun? Before we answer this we must try to get some idea of the size of this stupendous body. It is not the least use attempting to understand it by plain figures, for the figures would be too great to make any impression on us—they would be practically meaningless ; we must turn to some other method. Suppose, for instance, that the sun were a hollow ball ; then, if the earth were set at the centre, the moon could revolve round her at the same distance she is now, and there would be as great a distance between the moon and the shell of the sun as there is between the moon and the earth. This gives us a little idea of the size of the sun. Again, if we go back to that solar system in which we represented the planets by various objects from a pea to a football, and set a lamp in the centre to do duty for the sun, what size do you suppose that lamp would have to be really to represent the sun in proportion to the planets ? Well, if our greengage plum which did duty for the earth were about three-quarters of an inch in diameter we should want a lamp with a flame as tall as the tallest man you know, and even then it would not give a correct idea unless you imagined that man extending his arms widely, and you drew round him a circle and filled in all the circle with flame ! If this glorious flame burnt clear and fair and bright, radiating beams of light all around, the little greengage plum would not have to be too near, or it would be shrivelled up as in the blast of a furnace. To place it at anything resembling the distance it is from the sun in reality you would have to walk away from the flaming light for about three hundred steps, and set it down there; then, after having done all this, you would have some little idea of the relative sizes of the sun and the earth, and of the distance between them.

Of course, all the other planets would have to be at corresponding distances. On this same scale, Neptune, the furthest out, would be three miles from our artificial sun! It seems preposterous to think that some specks so small as to be quite invisible, specks that crawl about on that plum, have dared to weigh and measure the gigantic sun; but yet they have done it, and they have even decided what he is made of. The result of the experiments is that we know the sun to be a ball of glowing gas at a temperature so high that nothing we have on earth could even compare with it. Of his radiating beams extending in all directions few indeed fall on our little plum, but those that do are the source of all life, whether animal or vegetable. If the sun's rays were cut off from us, we should die at once. Even the coal we use to keep us warm is but sun's heat stored up ages ago, when the luxuriant tropical vegetation sprang up in the warmth and then fell down and was buried in the earth.   At night we are still enjoying the benefit of the sun's rays—that is, of those which are retained by our atmosphere ; for if none remained even the very air itself would freeze, and by the next morning not one inhabitant would be left alive to tell the awful tale.   Yet all this life and growth and heat we receive on the whole earth is but one part in two thousand two hundred millions of parts that go out in all directions into space.   It has been calculated that the heat which falls on to all the planets together cannot be more than one part in one hundred millions and the other millions of parts seem to us to be simply wasted.

For untold ages the sun has been pouring out this prodigal profusion of glory, and as we know that this cannot go on without some sort of compensation, we want to understand what keeps up the fires in the sun.   It is true that the sun is so enormous that he might go on burning for a very long time without burning right away ; but, then, even if he is huge, his expenditure is also huge.   If he had been made of solid coal he would have been all used up in about six thousand years, burning at the pace he does.   Now, we know that the ancient Egyptians kept careful note of the heavenly bodies, and if the sun were really burning away he must have been very much larger in their time ; but we have no record of this ; on the contrary, all records of the sun even to five thousand years ago show that he was much the same as at present. It is evident that we must search elsewhere for an explanation. It has been suggested that his furnace is supplied by the number of meteors that fall into him. Meteors are small bodies of the same materials as the planets, and may be likened to the dust of the solar system. It is not difficult to calculate the amount of matter he would require on this assumption to keep him going, and the amount required is so great as to make it practically impossible that this is the source of his supply. We have seen that all matter influences all other matter, and the quantity of meteoric stuff that would be required to support the sun's expenditure would be enough to have a serious effect on Mercury, an effect that would certainly have been noticed. There can, therefore, be no such mass of matter near the sun, and though there is no doubt a certain number of meteors do fall into his furnaces day by day, it is not nearly enough to account for his continuous radiation. It seems after this as if nothing else could be suggested ; but yet an answer has been found, an answer so wonderful that it is more like a fairy tale than reality.

To begin at the beginning, we must go back to the time when the sun was only a great gaseous nebula filling all the space included in the orbit of Neptune. This nebula was not in itself hot, but as it rotated it contracted. Now, heat is really only a form of energy, and energy and heat can be interchanged easily. This is a very startling thing when heard for the first time, but it is known as surely as we know anything and has been proved again and again. When a savage wants to make a fire he turns a piece of hard wood very very quickly between his palms—twiddles it, we should say expressively— into a hole in another piece of wood, until a spark bursts out. What is the spark? It is the energy of the savage's work turned to heat. When a horse strikes his iron-shod hoofs hard on the pavement you see sparks fly ; that is caused by the energy of the horse's leg. When you pump hard at your bicycle you feel your pump getting quite hot, for part of the energy you are putting into your work is transformed into heat ; and so on in numberless instances. No energetic action of any kind in this world takes place without some of the energy being turned into heat, though in many instances the

amount is so small as to be unnoticeable.   Nothing
falls to the ground without some heat being
generated.   Now, when this great nebula first began
its remarkable career, by the action of gravity all
the particles in it were drawn toward the centre ;
little by little they fell in, and the nebula became
smaller.   We are not now concerned with the
origin of the planets—we leave that aside ; we are
only contemplating the part of the nebula which
remained to become the sun.   Now these particles
being drawn inward each generated some heat,
so as the nebula contracted its temperature rose.
Throughout the ages, over the space of millions and
millions of miles, it contracted and grew hotter.   It
still remained gaseous, but at last it got to an
immense temperature, and is the sun as we know it.
What then keeps it shining ?   It is still contracting,
but slowly, so slowly that it is quite imperceptible
to our finest instruments.   It has been calculated
that if it contracts two hundred and fifty feet in
diameter in a year, the energy thus gained and
turned into heat is quite sufficient to account for its
whole yearly output.   This is indeed marvellous.
In comparison with the sun's size two hundred and
fifty feet is nothing.   It would take nine thousand
years at this rate before any diminution could be

noticed by our finest instruments! Here is a source
of heat which can continue for countless ages without exhaustion. Thus to all intents and purposes
we may say the sun's shining is inexhaustible.
Yet we must follow out the train of reasoning, and
see what will happen in the end, in eras and eras of
time, if nothing intervenes. Well, some gaseous
bodies are far finer and more tenuous than others,
and when a gaseous body contracts it is all the time
getting denser ; as it grows denser and denser it at
last becomes liquid, and then solid, and then it
ceases to contract, as of course the particles of a
solid body cannot fall freely toward the centre, as
those of a gaseous body can. Our earth has long
ago reached this stage. When solid the action
ceases, and the heat is no more kept up by this
source of energy, therefore the body begins to cool—
surface first, and lastly the interior ; it cools more
quickly the smaller it is. Our moon has parted
with all her heat long ago, while the earth still
retains some internally. In the sun, therefore, we
have an object-lesson of the stages through which all
the planets must have passed. They have all once
been glowing hot, and some may be still hot even
on the surface, as we have seen there is reason to
believe is the case with Jupiter.

By this marvellous arrangement for the continued heat of the sun we can see that the warmth of our planets is assured for untold ages. There is no need to fear that the sun will wear out by burning. His brightness will continue for ages beyond the thoughts of man.

Besides this, a few other things have been discovered about him. He is, of course, exceptionally difficult to observe; for though he is so large, which should make it easy, he is so brilliant that anyone regarding him through a telescope without the precaution of prepared glasses to keep off a great part of the light would be blinded at once. One most remarkable fact about the sun is that his surface is flecked with spots, which appear sometimes in greater numbers and sometimes in less, and the reason and shape of these spots have greatly exercised men's minds. Sometimes they are large enough to be seen without a telescope at all, merely by looking through a piece of smoked or coloured glass, which cuts off the most overpowering rays. When they are visible like this they are enormous, large enough to swallow many earths in their depths. At other times they may be observed by the telescope, then they may be about five thousand miles across. Sometimes one spot can be followed by an astronomer as it passes all across the sun, disappears at the edge, and after a lapse of time comes back again round the other edge. This first showed men that the sun, like all the planets, rotated on his axis, and gave them the means of finding out how long he took in doing so. But the spots showed a most surprising result, for they took slightly different times in making their journey round the sun, times which differed according to their position. For instance, a spot near the equator of the sun took twenty-five days to make the circuit, while one higher up or lower down took twenty-six days, and one further out twenty-seven; so that if these spots are, as certainly believed, actually on the surface, the conclusion is that the sun does not rotate all in one piece, but that some parts go faster than others. No one can really explain how this could be, but it is certainly more easily understood in the case of a body of gas than of a solid body, when it would be simply impossible to conceive. The spots seem to keep principally a little north and a little south of the equator; there are very few actually at it, and none found near the poles, but no reason for this distribution has been discovered. It has been noted that about every eleven years the greatest number of spots appears,

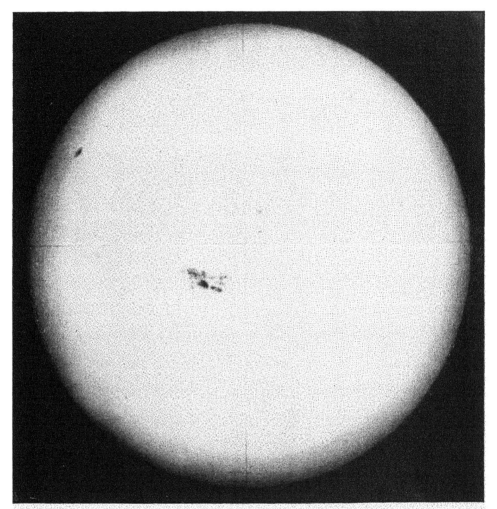

SUN-SPOTS.

Royal Observatory, Greenwich.

and that they become fewer again, mounting up in number to the next eleven years, and so on. All these curious facts show there is much yet to be solved about the sun. The spots were supposed for long to be eruptions bursting up above the surface, but now they are generally held to be deep depressions like saucers, probably caused by violent tempests, and it is thought that the inrush of cooler matter from above makes them look darker than the other parts of the sun's surface. But when we use the words 'cooler' and 'darker,' we mean only by comparison, for in reality the dark parts of the spots are brighter than electric light.

The fact that the spots are in reality depressions or holes is shown by their change of appearance as they pass over the face of the sun toward the edge ; for the change of shape is exactly that which would be caused by foreshortening.

It sounds odd to say that the best time for observing the sun is during a total eclipse, for then the sun's body is hidden by the moon. But yet to a certain extent this is true, and the reason is that the sun's own brilliance is our greatest hindrance in observing him, his rays are so dazzling that they light up our own atmosphere, which prevents us seeing the edges. Now, during a total eclipse,

7—2

when nearly all the rays are cut off, we can see marvellous things, which are invisible at other times. But total eclipses are few and far between, and so when one is approaching astronomers make great preparations beforehand.

A total eclipse is not visible from all parts of the world, but only from that small part on which the shadow of the moon falls, and as the earth travels, this shadow, which is really a round spot, passes along, making a dark band. In this band astronomers choose the best observatories, and there they take up their stations. The dark body of the moon first appears to cut a little piece out of the side of the sun, and as it sails on, gradually blotting out more and more, eager telescopes follow it ; at last it covers up the whole sun, and then a marvellous spectacle appears, for all round the edges of the black moon are seen glorious red streamers and arches and filaments of marvellous shapes, continually changing. These are thrown against a backgrouud of pale green light that surrounds the black moon and the hidden sun. In early days astronomers thought these wonderful coloured streamers belonged to the moon ; but it was soon proved that they really are part of the sun, and are only invisible at ordinary times, because our atmosphere is too bright to allow them to be seen. An instrument has now been invented to cut off most of the light of the sun, and when this is attached to a telescope these prominences, as they are called, can be seen at any time, so that there is no need to wait for an eclipse.

What are these marvellous streamers and filaments ? They are what they seem, eruptions of fiery matter discharged from the ever-palpitating sun thousands of miles into surrounding space. They are for ever shooting out and bursting and falling back, fireworks on a scale too enormous for us to conceive. Some of these brilliant flames extend for three hundred thousand miles, so that in comparison with one of them the whole world would be but a tiny ball, and this is going on day and night without cessation. Look at the picture where the artist has made a little black ball to represent the earth as she would appear if she could be seen in the midst of the flames shooting out from the sun. Do not make a mistake and think the earth really could be in this position ; she is only shown there so that you may see how tiny she is in comparison with the sun. All the time you have lived and your father, and grandfather, and right back to the beginnings of English history, and far, far further into the dim ages, this stupendous exhibition of energy and power has continued, and only of late years has anyone known anything about it ; even now a mere handful of people do know, and the rest, who are warmed and fed and kept alive by the gracious beams of this great revolving glowing fireball, never give it a thought.

I said just now a pale green halo surrounded the sun, extending far beyond the prominences ; this is called the corona and can only be seen during an eclipse. It surrounds the sun in a kind of shell, and there is reason to believe that it too is made of luminous stuff ejected by the sun in its burning fury. It is composed of large streamers or filaments, which seem to shoot out in all directions ; generally these are not much larger than the apparent width of the sun, but sometimes they extend much further. The puzzle is, this corona cannot be an atmosphere in any way resembling that of our earth ; for the gravitational force of the sun, owing to its enormous size, is so great that it would make any such atmosphere cling to it much more densely near to the surface, while it would be thinner higher up, and the corona is not dense in any way, but thin and tenuous throughout. This makes it very difficult to explain ; it is supposed that some kind of electrical force enters into the problem, but what it is exactly we are far from knowing yet.

CHAPTER VIII

SHINING VISITORS

OUR solar system is set by itself in the midst of a great space, and so far as we have learnt about it in this book everything in it seems orderly: the planets go round the sun and the satellites go round the planets, in orbits more or less regular; there seems no place for anything else. But when we have considered the planets and the satellites, we have not exhausted all the bodies which own allegiance to the sun. There is another class, made up of strange and weird members, which flash in and out of the system, coming and going in all directions and at all times—sometimes appearing without warning, sometimes returning with a certain regularity, sometimes retiring to infinite depths of space, where no human eye will ever see them more. These strange visitors are called comets, and are of all shapes and sizes and never twice alike. Even as we watch them they grow and change, and then diminish in splendour. Some are so vast that men see them as flaming signs in the sky, and regard them with awe and wonder ; some cannot be seen at all without the help of the telescope. From the very earliest ages those that were large enough to be seen without glasses have been regarded with astonishment. Men used to think that they were signs from heaven foretelling great events in the world. Timid people predicted that the end of the world would come by collision with one of them. Others, again, fancifully likened them to fishes in that sea of space in which we swim—fishes gigantic and terrifying, endowed with sense and will.

It is perhaps unnecessary to say that comets are no more alive than is our own earth, and as for causing the end of the world by collision, there is every reason to believe the earth has been more than once right through a comet's tail, and yet no one except scientific men even discovered it. These mysterious visitors from the outer regions of space were called comets from a Greek word signifying hair, for they often leave a long luminous trail behind, which resembles the filaments of a woman's hair. It is not often that one appears large and bright enough to be seen by the naked eye, and when it does it is not likely to be soon forgotten. In the year 1910 such a comet is expected, a comet which at its former appearance compelled universal attention by its brilliancy and strangeness. At the time of the Norman Conquest of England a comet believed to be the very same one was stretching its glorious tail half across the sky, and the Normans seeing it, took it as a good omen, fancying that it foretold their success. The history of the Norman Conquest was worked in tapestry—that is to say, in what we should call crewels on a strip of linen—and in this record the comet duly appears. Look at it in the picture as the Normans fancied it. It has a red head with blue flames starting from it, and several tails. The little group of men on the left are pointing and chattering about it. We can judge what an impression this comet must have made to be recorded in such an important piece of work.

But we are getting on too fast. We have yet to learn how anyone can know that the comet which appeared at the time of the Norman Conquest is the same as that which has come back again at different times, and above all, how anyone can tell that it will come again in the year 1910. All this involves a long story.

Before the invention of telescopes of course only those comets could be seen which were of great size and fine appearance.   In those days men did not realize that our world was but one of a number and of no great importance except to ourselves, and they always took these blazing appearances in the heavens as a particular warning to the human race. But when astronomers, by the aid of the telescope, found that for one comet seen by the eye there were hundreds which no mortal eye unaided could see, this idea seemed, to say the least of it, unlikely. Yet even then comets were looked upon as capricious visitors from outer space ; odd creatures drawn into our system by the attraction of the sun, who disappeared, never to return.   It was Newton, the same genius who disclosed to us the laws of gravity, who first declared that comets moved in orbits, only that these orbits were far more erratic than any of those followed by the planets.

So far we have supposed that the planets were all on what we should call a level—that is to say, we have regarded them as if they were floating in a sea of water around the sun ; but this is only approximately correct, for the orbits of the planets are not all at one level.   If you had a number of slender hoops or rings to represent the planetary orbits, you would have to tilt one a little this way and another a little that way, only never so far but that a line through the centre of the hoop from one side to another could pass through the sun. The way in which the planetary orbits are tilted is slight in comparison with that of the orbits of comets, for these are at all sorts of angles—some turned almost sideways, and others slanting, and all of them are ellipses long drawn out and much more irregular than the planetary orbits; but erratic as they are, in every case a line drawn through the sun and extended both ways would touch each side of the orbits.

A great astronomer called Halley, who was born in the time of the Commonwealth, was lucky enough to see a very brilliant comet, and the sight interested him so much that he made all the calculations necessary to find out just in what direction it was travelling in the heavens. He found out that it followed an ellipse which brought it very near to the sun at one part of its journey, and carried it far beyond the orbit of the earth, right out to that of Neptune, at the other. Then he began to search the records for other comets which had been observed before his time. He found that two particularly bright ones had been carefully noted—one about seventy-five years before that which he had seen, and the other seventy-five years before that again. Both these comets had been watched so scientifically that the paths in which they had travelled could be computed. A brilliant inspiration came to Halley. He believed that instead of these three, his own and the other two, being different comets, they were the same one, which returned to the sun about every seventy-five years. This could be proved, for if this idea were correct, of course the comet would return again in another seventy-five years, unless something unforeseen occurred. But Halley was in the prime of life: he could not hope to live to see his forecast verified. The only thing he could do was to note down exact particulars, by means of which others who lived after him might recognize his comet. And so when the time came for its return, though Halley was in his grave, numbers of astronomers were watching eagerly to see the fulfilment of his prediction. The comet did indeed appear, and since then it has been seen once again, and now we expect it to come back in the year 1910, when you and I may see it for ourselves. When the identity of the comet was fully established men began to search further back still, to compare the records of other previous brilliant comets, and found that this one had been noticed many times before, and once as I said, at the time of the Norman Conquest. Halley's comet is peculiar in many ways. For instance, it is unusual that so large and interesting a comet should return within a comparatively limited time. It is the smaller comets, those that can only be seen telescopically, that usually run in small orbits. The smallest orbits take about three and a half years to traverse, and some of the largest orbits known require a period of one hundred and ten thousand years. Between these two limits lies every possible variety of period. One comet, seen about the time Napoleon was born, was calculated to take two thousand years to complete its journey, and another, a very brilliant one seen in 1882, must journey for eight hundred years before it again comes near to the sun. But we never know what might happen, for at any moment a comet which has traversed a long solitary pathway in outer darkness may flash suddenly into our ken, and be for the first time noted and recorded, before flying off at an angle which must take it for ever further and further from the sun.

Everything connected with comets is mysterious and most fascinating. From out of the icy regions of space a body appears ; what it is we know not, but it is seen at first as a hairy or softly-glowing star, and it was thus that Herschel mistook Uranus for a comet when he first discovered it. As it draws nearer the comet sends out some fan-like projections toward the sun, enclosing its nucleus in filmy wrappings like a cocoon of light, and it travels faster and faster. From its head shoots out a tail—it may be more than one—growing in splendour and width, and always pointing away from the sun. So enormous are some of these tails that when the comet's head is close to the sun the tail extends far beyond the orbit of the earth. Faster still and faster flies the comet, for as we have seen it is a consequence of the law of gravitation that the nearer planets are to the sun the faster they move in their orbits, and the same rule applies to comets too. As the comet dashes up to the sun his pace becomes something indescribable; it has been reckoned for some comets at three hundred miles a second! But behold, as the head flies round the sun the tail is always projected outwards. The nucleus or head may be so near to the sun that the heat it receives would be sufficient to reduce molten iron to vapour; but this does not seem to affect it: only the tail expands. Sometimes it becomes two or more tails, and as it sweeps round behind the head it has to cover a much greater space in the same time, and therefore it must travel even faster than the head. The pace is such that no calculations can account for it, if the tail is composed of matter in any sense as we know it. Then when the sun is passed the comet sinks away again, and as it goes the tail dies down and finally disappears. The comet itself dwindles to a hairy star once more and goes—whither? Into space so remote that we cannot even dream of it—far away into cold more appalling than anything we could measure, the cold of absolute space. More and more slowly it travels, always away and away, until the sun, a short time back a huge furnace covering all the sky, is now but a faint star. Thus on its lonely journey unseen and unknown the comet goes.

This comet which we have taken as an illustration is a typical one, but all are not the same. Some have no tails at all, and never develop any; some change utterly even as they are watched. The same comet is so different at different times that the only possible way of identifying it is by knowing its path, and even this is not a certain method, for some comets appear to travel at intervals along the same path!

Now we come to the question that must have been in the mind of everyone from the beginning of this chapter, What are comets? This question no one can answer definitely, for there are many things so puzzling about these strange appearances that it is difficult even to suggest an explanation. Yet a good deal is known. In the first place, we are certain that comets have very little density— that is to say, they are indescribably thin, thinner than the thinnest kind of gas; and air, which we always think so thin, would be almost like a blanket compared with the material of comets. This we judge because they exercise no sort of influence on any of the planetary bodies they draw near to, which they certainly would do if they were made of any kind of solid matter. They come sometimes very close to some of the planets. A comet was so near to Jupiter that it was actually in among his moons. The comet was violently agitated; he was pulled in fact right out of his old path, and has been going on a new one ever since; but he did not exercise the smallest effect on Jupiter, or even on the moons. And, as I said earlier in this chapter, we on the earth have been actually in the folds of a comet's tail. This astonishing fact happened in June, 1861. One evening after the sun had set a golden-yellow disc, surrounded with filmy wrappings, appeared in the sky. The sun's light, diffused throughout our atmosphere, had prevented its being seen sooner. This was apparently the comet's head. It is described as 'though a number of light, hazy clouds were floating around a miniature full moon.' From this a cone of light extended far up into the sky, and when the head disappeared below the horizon this tail was seen to reach to the zenith. But that was not all. Strange shafts of light seemed to hang right overhead, and could only be accounted for by supposing that they were caused by another tail hanging straight above us, so that we looked up at it foreshortened by perspective. The comet's head lay between the earth and the sun, and its tail, which extended over many millions of miles, stretched out behind in such a way that the earth must have gone right through it. The fact that the comet exercised no perceptible influence on the earth at all, and that there were not even any unaccountable magnetic storms or displays of electricity, may reassure us so that if ever we do again come in contact with one of these extremely fine, thin bodies, we need not be afraid.

There is another way in which we can judge of the wonderful tenuity or thinness of comets—that

8

is, that the smallest stars can be seen through their tails, even though those tails must be many thousands of miles in thickness. Now, if the tails were anything approaching the density of our own atmosphere, the stars when seen through them would appear to be moved out of their places. This sounds odd, and requires a word of explanation. The fact is that anything seen through any transparent medium like water or air is what is called refracted—that is to say, the rays coming from it look bent. Everyone is quite familiar with this in everyday life, though perhaps they may not have noticed it. You cannot thrust a stick into the water without seeing that it looks crooked. Air being less dense than water has not quite so strong a refracting power, but still it has some. We cannot prove it in just the same way, because we are all inside the atmosphere ourselves, and there is no possibility of thrusting a stick into it from the outside! The only way we know it is by looking at something which is 'outside' already, and we find plenty of objects in the sky. As a matter of fact, the stars are all a little pulled out of their places by being seen through the air, and though of course we do not notice this, astronomers know it and have to make allowance for it. The effect

A STICK THRUST INTO THE WATER APPEARS CROOKED.

is most noticeable in the case of the sun when he is going down, for the atmosphere bends his rays up, and though we see him a great glowing red ball on the horizon, and watch him, as we think, drop gradually out of sight, we are really looking at him for the last moment or two when he has already gone, for the rays are bent up by the air and his image lingers when the real sun has disappeared.

Therefore in looking through the luminous stuff that forms a comet's tail astronomers might well expect to see the stars displaced, but not a sign of this appears. It is difficult to imagine, therefore, what the tail can be made of. The idea is that the sun exercises a sort of repulsive effect on certain elements found in the comet's head—that is to say, it pushes them away, and that as the head approaches the sun, these elements are driven out of it away from the sun in vapour. This action may have something to do with electricity, which is yet little understood; anyway, the effect is that, instead of attracting the matter toward itself, in which case we should see the comet's tails stretching toward the sun, the sun drives it away! In the chapter on the sun we had to imagine something of the same kind to account for the corona, and the corona and the comet's tails may be really akin to each other, and could perhaps be explained in the same way. Now we come to a stranger fact still. Some comets go right through the sun's corona, and yet do not seem to be influenced by it in the smallest degree. This may not seem very wonderful at first perhaps, but if you remember that a dash through anything so dense as our atmosphere, at a pace much less than that at which a comet goes, is enough to heat iron to a white heat, and then make it fly off in vapour, we get a glimpse of the extreme fineness of the materials which make the corona.

Here is Herschel's account of a comet that went very near the sun:

'The comet's distance from the sun's centre was about the 160th part of our distance from it. All the heat we enjoy on this earth comes from the sun. Imagine the heat we should have to endure if the sun were to approach us, or we the sun, to one 160th part of its present distance. It would not be merely as if 160 suns were shining on us all at once, but, 160 times 160, according to a rule which is well known to all who are conversant with such matters. Now, that is 25,600. Only imagine a glare 25,600 times fiercer than that of the equatorial sunshine at noon day with the sun vertical. In such a heat there is no substance we know of which would not run like water, boil, and be converted into smoke or vapour. No wonder the comet gave evidence of violent excitement, coming from the cold region outside the planetary system torpid and ice-bound. Already when arrived even in our temperate region it began to show signs of internal activity ; the head had begun to develop, and the tail to elongate, till the comet was for a time lost sight of—not for days afterwards was it seen ; and its tail, whose direction was reversed, and which could not possibly be the same tail it had before, had already lengthened to an extent of about ninety millions of miles, so that it must have been shot out with immense force in a direction away from the sun.'

We remember that comets have sometimes more than one tail, and a theory has been advanced to account for this too. It is supposed that perhaps different elements are thrust away by the sun at different angles, and one tail may be due to one element and another to another. But if the comet goes on tail-making to a large extent every time it returns to the sun, what happens eventually ? Do the tails fall back again into the head when out of reach of the sun's action ? Such an idea is inconceivable ; but if not, then every time a comet approaches the sun it loses something, and that something is made up of the elements which were formerly in the head and have been violently ejected. If this be so we may well expect to see comets which have returned many times to the sun without tails at all, for all the tail-making stuff that was in the head will have been used up, and as this is exactly what we do see, the theory is probably true.

Where do the comets come from? That also is a very large question. It used to be supposed they were merely wanderers in space who happened to have been attracted by our sun and drawn into his system, but there are facts which go very strongly against this, and astronomers now generally believe that comets really belong to the solar system, that their proper orbits are ellipses, and that in the case of those which fly off at such a speed that they can never return they must at some time have been pulled out of their original orbit by the influence of one of the planets.

To get a good idea of a really fine comet, until we have the opportunity of seeing one for ourselves, we cannot do better than look at this picture of a comet photographed in 1901 at the Cape of Good Hope. It is only comparatively recently that photography has been applied to comets. When

Royal Observatory, Cape of Good Hope

A GREAT COMET.

Halley's comet appeared last time such a thing was not thought of, but when he comes again numbers of cameras, fitted up with all the latest scientific appliances, will be waiting to get good impressions of him.

CHAPTER IX

SHOOTING STARS AND FIERY BALLS

ALL the substances which we are accustomed to see and handle in our daily lives belong to our world. There are vegetables which grow in the earth, minerals which are dug out of it, and elementary things, such as air and water, which have always made up a part of this planet since man knew it. These are obvious, but there are other things not quite so obvious which also help to form our world. Among these we may class all the elements known to chemists, many of which have difficult names, such as oxygen and hydrogen. These two are the elements which make up water, and oxygen is an important element in air, which has nitrogen in it too. There are numbers and numbers of other elements perfectly familiar to chemists, of which many people never even hear the names. We live in the midst of these things, and we take them for granted and pay little attention to them; but when we begin to learn about other worlds we at once want to know if these substances and elements which enter so largely into our daily lives are to be found elsewhere in the universe or are quite peculiar to our own world. This question might be answered in several ways, but one of the most practical tests would be if we could get hold of something which had not been always on the earth, but had fallen upon it from space. Then, if this body were made up of elements corresponding with those we find here, we might judge that these elements are very generally diffused throughout the bodies in the solar system.

It sounds in the highest degree improbable that anything should come hurling through the air and alight on our little planet, which we know is a mere speck in a great ocean of space; but we must not forget that the power of gravity increases the chances greatly, for anything coming within a certain range of the earth, anything small enough, that is, and not travelling at too great a pace, is bound to fall on to it. And, however improbable it seems, it is undoubtedly true that masses of matter do crash down upon the earth from time to time, and these are called meteorites. When we think of the great expanse of the oceans, of the ice round the poles, and of the desert wastes, we know that for every one of such bodies seen to fall many more must have fallen unseen by any human being. Meteors large enough to reach the earth are not very frequent, which is perhaps as well, and as yet there is no record of anyone's having been killed by them. Most of them consist of masses of stone, and a few are of iron, while various substances resembling those that we know here have been found in them. Chemists in analyzing them have also come across certain elements so far unknown upon earth, though of course there is no saying that these may not exist at depths to which man has not penetrated.

A really large meteor is a grand sight. If it is seen at night it appears as a red star, growing rapidly bigger and leaving a trail of luminous vapour behind as it passes across the sky. In the daytime this vapour looks like a cloud. As the meteor hurls itself along there may be a deep continuous roar, ending in one supreme explosion, or perhaps in several explosions, and finally the meteor may come to the earth in one mass, with a force so great that it buries itself some feet deep in the soil, or it may burst into numbers of tiny fragments, which are scattered over a large area.

When a meteor is found soon after its fall it is very hot, and all its surface has 'run,' having been fused by heat. The heat is caused by the friction of our atmosphere. The meteor gets entangled in the atmosphere, and, being drawn by the attraction of the earth, dashes through it. Part of the energy of its motion is turned to heat, which grows greater and greater as the denser air nearer to the earth is encountered; so that in time all the surface of the meteor runs like liquid, and this liquid, rising to a still higher temperature, is blown off in vapour, leaving a new surface exposed. The vapour makes the trail of fire or cloud seen to follow the meteor. If the process went on for long the meteor would be all dissipated in vapour, and in any case it must reach the earth considerably reduced in size.

Numbers and numbers of comparatively small ones disappear, and for every one that manages to come to earth there must be hundreds seen only as shooting stars, which vanish and 'leave not a wrack behind.' When a meteor is seen to fall it is traced, and, whenever possible, it is found and placed in a museum. Men have sometimes come across large masses of stone and iron with their surfaces fused with heat. These are in every way like the recognized meteorites, except that no eye has noted their advent. As there can be no reasonable doubt that they are of the same origin as the others, they too are collected and placed in museums, and in any large museum you would be able to see both kinds—those which have been seen to come to earth and those which have been found accidentally.

The meteors which appear very brilliant in their course across the sky are sometimes called fire-balls, which is only another name for the same thing. Some of these are brighter than the full moon, so bright that they cause objects on earth to cast a shadow. In 1803 a fiery ball was noticed above a small town in Normandy; it burst and scattered stones far and wide, but luckily no one was hurt. The largest meteorites that have been found on the earth are a ton or more in weight; others are mere stones; and others again just dust that floats about in the atmosphere before gently settling. Of course, meteors of this last kind could not be seen to fall like the larger ones, yet they do fall in such numbers that calculations have been made showing that the earth must catch about a hundred millions of meteors daily, having altogether a total weight of about a hundred tons. This sounds enormous, but compared with the weight of the earth it is very small indeed.

Now that we have arrived at the fact that strange bodies do come hurtling down upon us out of space, and that we can actually handle and examine them, the next question is, Where do they come from ? At one time it was thought that they were fragments which had been flung off by the earth herself when she was subject to violent explosions, and that they had been thrown far enough to resist the impulse to drop down upon her again, and had been circling round the sun ever since, until the earth came in contact with them again and they had fallen back upon her. It is not difficult to imagine a force which would be powerful enough to achieve the feat of speeding something off at such a velocity that it passed beyond the earth's power to pull it back, but nothing that we have on earth would be nearly strong enough to achieve such a feat. Imaginative writers have pictured a projectile hurled from a cannon's mouth with such tremendous force that it not only passed beyond the range of the earth's power to pull it back, but so that it fell within the influence of the moon and was precipitated on to her surface ! Such things must remain achievements in imagination only ; it is not possible for them to be carried out.   Other ideas as to the origin of meteors were that they had been expelled from the moon or from the sun.   It would need a much less force to send a projectile away from the moon than from the earth on account of its smaller size and less density, but the distance from the earth to the moon is not very great, and any projectile hurled forth from the moon would cross it in a comparatively short time. Therefore if the meteorites come from the moon, the moon must be expelling them still, and we might expect to see some evidence of it; but we know that the moon is a dead world, so this explanation is not possible.   The sun, for its part, is torn by such gigantic disturbances that, notwithstanding its vast size, there is no doubt sufficient force there to send meteors even so far as the earth, but the chances of their encountering the earth would be small.   Both these theories are now discarded.   It is believed that the meteors are merely lesser fragments of the same kind of materials as the planets, circling independently round the sun; and a proof of this is that far more meteorites fall on that part of the earth which is facing forward in its journey than on that behind, and this is what we should expect if the meteors were scattered independently through space and it was by reason of our movements that we came in contact with them. There is no need to explain this further. Everyone knows that in cycling or driving along a road where there is a good deal of traffic both ways the people we meet are more in number than those who overtake us, and the same result would follow with the meteors ; that is to say, in travelling through space where they were fairly evenly distributed we should meet more than we should be overtaken by.

You remember that it was suggested the sun's fuel might be obtained from meteors, and this was proved to be not possible, even though there are no doubt unknown millions of these strange bodies circling throughout the solar system.

There are so many names for these flashing bodies that we may get a little confused : when they are seen in the sky they are meteors, or fire-balls ; when they reach the earth they are called meteorites, and also aerolites. Then there is another class of the same bodies called shooting stars, and these are in reality only meteors on a smaller scale ; but there ought to be no confusion in our thoughts, for all these objects are small bodies travelling round the sun, and caught by the earth's influence.

When you watch the sky for some time on a

clear night, you will seldom fail to see at least one star flash out suddenly in a path of thrilling light and disappear, and you cannot be certain whether that star had been shining in the sky a minute before, or if it had appeared suddenly only in order to go out. The last idea is right. We must get rid at once of the notion that it would be possible for any fixed star to behave in this manner. To begin with, the fixed stars are many of them actually travelling at a great velocity at present, yet so immeasurably distant are they that their movement makes no perceptible difference to us. For one of them to appear to dash across the heavens as a meteor does would mean a velocity entirely unknown to us, even comparing it with the speed of light. No, these shooting stars are not stars at all, though they were so named, long before the real motions of the fixed stars were even dimly guessed at. As we have seen, they belong to the same class as meteors.

I remember being told by a clergyman, years ago, that one night in November he had gone up to bed very late, and as he pulled up his blind to look at the sky, to his amazement he saw a perfect hail of shooting stars, some appearing every minute, and all darting in vivid trails of light, longer or shorter,

though all seemed to come from one point.   So
marvellous was the sight that he dashed across the
village street, unlocked the church door, and himself
pulled the bell with all his might.   The people in
that quiet country village had long been in bed, but
they huddled on their clothes and ran out of their
pretty thatched cottages, thinking there must be
a great fire, and when they saw the wonder in
the sky they were amazed and cried out that the
world must be coming to an end.   The clergyman
knew better than that, and was able to reassure
them, and tell them he had only taken the most
effectual means of waking them so that they might
not miss the display, for he was sure as long as they
lived they would never see such another sight.   A
star shower of this kind is certainly well worth
getting up to see, but though uncommon it is not
unique.   There are many records of such showers
having occurred in times gone by, and when men
put together and examined the records they found
that the showers came at regular intervals.   For
instance, every year about the same time in
November there is a star shower, not comparable,
it is true, with the brilliant one the clergyman saw,
but still noticeable, for more shooting stars are seen
then than at other times, and once in every thirty-

three years there is a specially fine one. It happened in fact to be one of these that the village
people were wakened up to see.

Not all at once, but gradually, the mystery of
these shower displays was solved. It was realized
that the meteors need not necessarily come from
one fixed place in the sky because they seemed to
us to do so, for that was only an effect of perspective. If you were looking down a long,
perfectly straight avenue of tree-trunks, the avenue
would seem to close in, to get narrower and
narrower at the far end until it became a point;
but it would not really do so, for you would know
that the trees at the far end were just the same
distance from each other as those between which
you were standing. Now, two meteors starting
from the same direction at a distance from each
other, and keeping parallel, would seem to us to
start from a point and to open out wider and wider
as they approached, but they would not really do so;
it would only be, as in the case of the avenue, an effect
of perspective. If a great many meteors did the same
thing, they would appear to us all to start from one
point, whereas really they would be on parallel lines,
only as they rushed to meet us or we rushed to
meet them this effect would be produced. There-

fore the first discovery was that these meteors were
thousands and thousands of little bodies travelling
in lines parallel to each other, like a swarm of little
planets. To judge that their path was not a straight
line but a circle or ellipse was the next step, and
this was found to be the case. From taking exact
measurements of their paths in the sky an astronomer
computed they were really travelling round the sun
in a lengthened orbit, an ellipse more like a comet's
orbit than that of a planet. But next came the
puzzling question, Why did the earth apparently hit
them every year to some extent, and once in thirty-
three years seem to run right into the middle of
them ? This also was answered. One has only to
imagine a swarm of such meteors at first hastening
busily along their orbit, a great cluster all together,
then, by the near neighbourhood of some planet, or
by some other disturbing causes, being drawn out,
leaving stragglers lagging behind, until at last
thère might be some all round the path, but
only thinly scattered, while the busy, important
cluster that formed the nucleus was still much
thicker than any other part. Now, if the orbit
that the meteors followed cut the orbit or path
of the earth at one point, then every time the
earth came to what we may call the level crossing

she must run into some of the stragglers, and if the chief part of the swarm took thirty-three years to get round, then once in about thirty-three years the earth must strike right into it.   This would account for the wonderful display.   So long drawn-out is the thickest part of the swarm that it takes a year to pass the points at the level crossing.   If the earth strikes it near the front one year, she may come right round in time to strike into the rear part of the swarm next year, so that we may get fine displays two years running about every thirty-three years.   The last time we passed through the swarm was in 1899, and then the show was very disappointing.   Here in England thick clouds prevented our seeing much, and there will not be another chance for us to see it at its best until 1932.

These November meteors are called Leonids, because they *seem* to come from a group of stars named Leo, and though the most noticeable they are not the only ones.   A shower of the same kind occurs in August too, but the August meteors, called Perseids, because they seem to come from Perseus, revolve in an orbit which takes a hundred and forty-two years to traverse !   So that only every one hundred and forty-second year could we hope to see a good display.   When all these facts had been gathered up, it seemed without doubt that certain groups of meteors travelled in company along an elliptical orbit.   But there remained still something more—a bold and ingenious theory to be advanced.   It was found that a comet, a small one, only to be seen with the telescope, revolved in exactly the same orbit as the November meteors, and another one, larger, in exactly the same orbit as the August ones ; hence it could hardly be doubted that comets and meteors had some connection with each other, though what that connection is exactly no one knows.   Anyway, we can have no shadow of doubt when we find the comet following a marked path, and the meteors pursuing the same path in his wake, that the two have some mysterious affinity. There are other smaller showers besides these of November and August, and a remarkable fact is known about one of them.   This particular stream was found to be connected with a comet named Biela's Comet, that had been many times observed, and which returned about every seven years to the sun.   After it had been seen several times, this astonishing comet split in two and appeared as two comets, both of which returned at the end of the next seven years.   But on the next two occasions when they were expected they never came at all,

and the third time there came instead a fine display of shooting stars, so it really seemed as if
these meteors must be the fragments of the lost
comet.

It is very curious and interesting to notice that
in these star showers there is no certain record of
any large meteorite reaching the earth ; they seem
to be made up of such small bodies that they are
all dissipated in vapour as they traverse our air.

CHAPTER X

THE GLITTERING HEAVENS

On a clear moonless night the stars appear uncountable. You see them twinkling through the leafless trees, and covering all the sky from the zenith, the highest point above your head, down to the horizon. It seems as if someone had taken a gigantic pepper-pot and scattered them far and wide so that some had fallen in all directions. If you were asked to make a guess as to how many you can see at one time, no doubt you would answer 'Millions!' But you would be quite wrong, for the number of stars that can be seen at once without a telescope does not exceed two thousand, and this, after the large figures we have been dealing with, appears a mere trifle. With a telescope, even of small power, many more are revealed, and every increase in the size of the telescope shows more still; so that it might be supposed the universe is indeed illimitable, and that we are only prevented from seeing beyond a certain point by our limited

135

resources. But in reality we know that this cannot be so. If the whole sky were one mass of stars, as it must be if the number of them were infinite, then, even though we could not distinguish the separate items, we should see it bright with a pervading and diffused light. As this is not so, we judge that the universe is not unending, though, with all our inventions, we may never be able to probe to the end of it. We need not, indeed, cry for infinity, for the distances of the fixed stars from us are so immeasurable that to atoms like ourselves they may well seem unlimited. Our solar system is set by itself, like a little island in space, and far, far away on all sides are other great light-giving suns resembling our own more or less, but dwindled to the size of tiny stars, by reason of the great void of space lying between us and them. Our sun is, indeed, just a star, and by no means large compared with the average of the stars either. But, then, he is our own ; he is comparatively near to us, and so to us he appears magnificent and unique. Judging from the solar system, we might expect to find that these other great suns which we call stars have also planets circling round them, looking to them for light and heat as we do to our sun. There is no reason to doubt that in some instances the conjecture is right, and that there may be other suns with attendant planets. It is however a great mistake to suppose that because our particular family in the solar system is built on certain lines, all the other families must be made on the same pattern. Why, even in our own system we can see how very much the planets differ from each other: there are no two the same size; some have moons and some have not; Saturn's rings are quite peculiar to himself, and Uranus and Neptune indulge in strange vagaries. So why should we expect other systems to be less varied?

As science has advanced, the idea that these far-away suns must have planetary attendants as our sun has been discarded. The more we know the more is disclosed to us the infinite variety of the universe. For instance, so much accustomed are we to a yellow sun that we never think of the possibility of there being one of another colour. What would you say then to a ruby sun, or a blue one; or to two suns of different colours, perhaps red and green, circling round each other; or to two such suns each going round a dark companion? For there are dark bodies as well as shining bodies in the sky. These are some of the marvels of the starry sky, marvels quite as absorbing as anything we have found in the solar system.

It requires great care and patience and infinite labour before the very delicate observations which alone can reveal to us anything of the nature of the fixed stars can be accomplished. It is only since the improvement in large telescopes that this kind of work has become possible, and so it is but recently men have begun to study the stars intimately, and even now they are baffled by indescribable difficulties. One of these is our inability to tell the distance of a thing by merely looking at it unless we also know its size. On earth we are used to seeing things appear smaller the further they are from us, and by long habit can generally tell the real size ; but when we turn to the stars, which appear so much alike, how are we to judge how far off they are? Two stars apparently the same size and close together in the sky may really be as far one from another as the earth is from the nearest ; for if the further one were very much larger than the nearer, they would then appear the same size.

At first it was natural enough to suppose that the big bright stars of what we call the first magnitude were the nearest to us, and the less bright the

next nearest, and so on down to the tiny ones,
only revealed by the telescope, which would be
the furthest away of all; but research has shown
that this is not correct. Some of the brightest
stars may be comparatively near, and some of the
smallest may be near also. The size is no test
of distance. So far as we have been able to discover, the star which seems nearest *is* a first magnitude one, but some of the others which outshine it
must be among the infinitely distant ones. Thus
we lie in the centre of a jewelled universe, and
cannot tell even the size of the jewels which cover
its radiant robe.

I say 'lie,' but that is really not the correct word.
So far as we have been able to find out, there is no
such thing as absolute rest in the universe—in fact,
it is impossible; for even supposing any body could
be motionless at first, it would be drawn by the
attraction of its nearest neighbours in space, and
gradually gain a greater and greater velocity as it
fell toward them. Even the stars we call 'fixed'
are all hurrying along at a great pace, and though
their distance prevents us from seeing any change
in their positions, it can be measured by suitable
instruments. Our sun is no exception to this universal rule. Like all his compeers, he is hurrying

busily along somewhere in obedience to some impulse of which we do not know the nature; and as he goes he carries with him his whole cortège of planets and their satellites, and even the comets. Yes, we are racing through space with another motion, too, besides those of rotation and revolution, for our earth keeps up with its master attractor, the sun.   It is difficult, no doubt, to follow this, but if you think for a moment you will remember that when you are in a railway-carriage everything in that carriage is really travelling along with it, though it does not appear to move.   And the whole solar system may be looked at as if it were one block in movement.   As in a carriage, the different bodies in it continue their own movements all the time, while sharing in the common movement.   You can get up and change your seat in the train, and when you sit down again you have not only moved that little way of which you are conscious, but a great way of which you are not conscious unless you look out of the window. Now in the case of the earth's own motion we found it necessary to look for something which does not share in that motion for purposes of comparison, and we found that something in the sun, who shows us very clearly we are turning on our axis.

But in the case of the motion of the solar system
the sun is moving himself, so we have to look
beyond him again and turn to the stars for con-
firmation.   Then we find that the stars have
motions of their own, so that it is very difficult
to judge by them at all.   It is as if you were
bicycling swiftly towards a number of people all
walking about in different directions on a wide
lawn.   They have their movements, but they all
also have an apparent movement, really caused by
you as you advance toward them ; and what
astronomers had to do was to separate the true
movements of the stars from the false apparent
movement made by the advance of the sun.   This
great problem was attacked and overcome, and it is
now known with tolerable certainty that the sun is
sweeping onward at a pace of about twelve miles
a second toward a fixed point.   It really matters
very little to us where he is going, for the distances
are so vast that hundreds of years must elapse
before his movement makes the slightest difference
in regard to the stars.   But there is one thing
which we can judge, and that is that though his
course appears to be in a straight line, it is most
probably only a part of a great curve so huge that
the little bit we know seems straight.

When we speak of the stars, we ought to keep quite clearly in our minds the fact that they lie at such an incredible distance from us that it is probable we shall never learn a great deal about them. Why, men have not even yet been able to communicate with the planet Mars, at its nearest only some thirty-five million miles from us, and this is a mere nothing in measuring the space between us and the stars. To express the distances of the stars in figures is really a waste of time, so astronomers have invented another way. You know that light can go round the world eight times in a second; that is a speed quite beyond our comprehension, but we just accept it. Then think what a distance it could travel in an hour, in a day; and what about a year? The distance that light can travel in a year is taken as a convenient measure by astronomers for sounding the depths of space. Measured in this way light takes four years and four months to reach us from the nearest star we know of, and there are others so much more distant that hundreds—nay, thousands—of years would have to be used to convey it. Light which has been travelling along with a velocity quite beyond thought, silently, unresting, from the time when the Britons lived and ran half naked on this island of ours, has only reached us now, and there is no limit to the time we may go back in our imaginings. We see the stars, not as they are, but as they were. If some gigantic conflagration had happened a hundred years ago in one of them situated a hundred light-years away from us, only now would that messenger, swifter than any messenger we know, have brought the news of it to us. To put the matter in figures, we are sure that no star can lie nearer to us than twenty-five billions of miles. A billion is a million millions, and is represented by a figure with twelve noughts behind it, so—1,000,000,000,000 ; and twenty-five such billions is the least distance within which any star can lie. How much further away stars may be we know not, but it is something to have found out even that. On the same scale as that we took in our first example, we might express it thus : If the earth were a greengage plum at a distance of about three hundred of your steps from the sun, and Neptune were, on the same scale, about three miles away, the nearest fixed star could not be nearer than the distance measured round the whole earth at the Equator !

All this must provoke the question, How can anyone find out these things ? Well, for a long time the problem of the distances of the stars was thought to be too difficult for anyone to attempt to solve it, but at last an ingenious method was devised, a method which shows once more the triumph of man's mind over difficulties. In practice this method is extremely difficult to carry out, for it is complicated by so many other things which must be made allowance for; but in theory, roughly explained, it is not too hard for anyone to grasp. The way of it is this: If you hold up your finger so as to cover exactly some object a few feet distant from you, and shut first one eye and then the other, you will find that the finger has apparently shifted very considerably against the background. The finger has not really moved, but as seen from one eye or the other, it is thrown on a different part of the background, and so appears to jump; then if you draw two imaginary lines, one from each eye to the finger, and another between the two eyes, you will have made a triangle. Now, all of you who have done a little Euclid know that if you can ascertain the length of one side of a triangle, and the angles at each end of it, you can form the rest of the triangle; that is to say, you can tell the length of the other two sides. In this instance the base line, as it is called—that is to say the line lying between the two eyes—can easily be measured, and the angles at each end can be found by an instrument called a sextant, so that by simple calculation anyone could find out what distance the finger was from the eye. Now, some ingenious man decided to apply this method to the stars. He knew that it is only objects quite near to us that will appear to shift with so small a base line as that between the eyes, and that the further away anything is the longer must the base line be before it makes any difference. But this clever man thought that if he could only get a base line long enough he could easily compute the distance of the stars from the amount that they appeared to shift against their background. He knew that the longest base line he could get on earth would be about eight thousand miles, as that is the diameter of the earth from one side to the other; so he carefully observed a star from one end of this immense base line and then from the other, quite confident that this plan would answer. But what happened? After careful observations he discovered that no star moved at all with this base line, and that it must be ever so much longer in order to make any impression. Then indeed the case seemed hopeless, for here we are tied to the earth and we cannot get

away into space.   But the astronomer was nothing
daunted.   He knew that in its journey round the
sun the earth travels in an orbit which measures
about one hundred and eighty-five millions of miles
across, so he resolved to take observations of the
stars when the earth was at one side of this great
circle, and again, six months later, when she had
travelled to the other side.   Then indeed he would
have a magnificent base line, one of one hundred
and eighty-five millions of miles in length.   What
was the result ?   Even with this mighty line the
stars are found to be so distant that many do not
move at all, not even when measured with the
finest instruments, and others move, it may be, the
breadth of a hair at a distance of several feet !   But
even this delicate measure, a hair's-breadth, tells its
own tale ; it lays down a limit of twenty-five billion
miles within which no star can lie !

This system which I have explained to you is
called finding the star's parallax, and perhaps it
is easier to understand when we put it the other
way round and say that the hair's-breadth is what
the whole orbit of the earth would appear to have
shrunk to if it were seen from the distance of these
stars !

Many, many stars have now been examined, and of

them all our nearest neighbour seems to be a bright star seen in the Southern Hemisphere. It is in the constellation or star group called Centaurus, and is the brightest star in it. In order to designate the stars when it is necessary to refer to them, astronomers have invented a system. To only the very brightest are proper names attached; others are noted according to the degree of their brightness, and called after the letters of the Greek alphabet: alpha, beta, gamma, delta, etc. Our own word 'alphabet' comes, you know, from the first two letters of this Greek series. As this particular star is the brightest in the constellation Centaurus, it is called Alpha Centauri; and if ever you travel into the Southern Hemisphere and see it, you may greet it as our nearest neighbour in the starry universe, so far as we know at present.

CHAPTER XI

THE CONSTELLATIONS

FROM the very earliest times men have watched the stars, felt their mysterious influence, tried to discover what they were, and noted their rising and setting. They classified them into groups, called constellations, and gave such groups the names of figures and animals, according to the positions of the stars composing them. Some of these imaginary figures seem to us so wildly ridiculous that we cannot conceive how anyone could have gone so far out of their way as to invent them. But they have been long sanctioned by custom, so now, though we find it difficult to recognize in scattered groups of stars any likeness to a fish or a ram or a bear, we still call the constellations by their old names for convenience in referring to them.

Supposing the axis of the earth were quite upright, straight up and down in regard to the plane at which the earth goes round the sun, then

148

we should always see the same set of stars from the Northern and the same set of stars from the Southern Hemispheres all the year round. But as the axis is tilted slightly, we can, during our nights in the winter in the Northern Hemisphere, see more of the sky to the south than we can in the summer; and in the Southern Hemisphere just the reverse is the case, far more stars to the north can be seen in the winter than in the summer. But always, whether it is winter or summer, there is one fixed point in each hemisphere round which all the other stars seem to swing, and this is the point immediately over the North or the South Poles. There is, luckily, a bright star almost at the point at which the North Pole would seem to strike the sky were it infinitely lengthened. This is not one of the brightest stars in the sky, but quite bright enough to serve the purpose, and if we stand with our faces towards it, we can be sure we are looking due north. How can we discover this star for ourselves in the sky? Go out on any starlight night when the sky is clear, and see if you can find a very conspicuous set of seven stars called the Great Bear. I shall not describe the Great Bear, because all children ought to know it already, and if they don't, they can ask the first grown-up person they meet, and they will certainly be told. (See map.)

Having found the Great Bear, you have only to draw an imaginary line between the two last stars forming the square on the side away from the tail, and carry it on about three times as far as the distance between those two stars, and you will come straight to the Pole Star. The two stars in the Great Bear which help one to find it are called the Pointers, because they point to it.

The Great Bear is one of the constellations known from the oldest times; it is also sometimes called Charles's Wain, the Dipper, or the Plough. It is always easily seen in England, and seems to swing round the Pole Star as if held by an invisible rope tied to the Pointers. Besides the Great Bear there is, not far from it, the Little Bear, which is really very like it, only smaller and harder to find. The Pole Star is the last star in its tail; from it two small stars lead away parallel to the Great Bear, and they bring the eye to a small pair which form one side of a square just like that in the Great Bear. But the whole of the Little Bear is turned the opposite way from the Great Bear, and the tail points in the opposite direction. And when you come to think of it, it

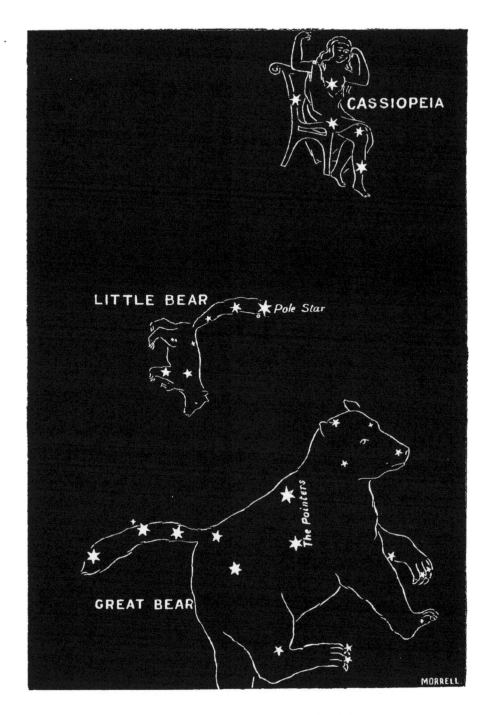

CONSTELLATIONS NEAR THE POLE STAR.

is very ridiculous to have called these groups Bears at all, or to talk about tails, for bears have no tails ! So it would have been better to have called them foxes or dogs, or almost any other animal rather than bears.

Now, if you look at the sky on the opposite side of the Pole Star from the Great Bear, you will see a clearly marked capital W made up of five or six bright stars. This is called Cassiopeia, or the Lady's Chair.

In looking at Cassiopeia you cannot help noticing that there is a zone or broad band of very many stars, some exceedingly small, which apparently runs right across the sky like a ragged hoop, and Cassiopeia seems to be set in or on it. This band is called the Milky Way, and crosses not only our northern sky, but the southern sky too, thus making a broad girdle round the whole universe. It is very wonderful, and no one has yet been able to explain it. The belt is not uniform and even, but it is here and there broken up into streamers and chips, having the same appearance as a piece of ribbon which has been snipped about by scissors in pure mischief; or it may be compared to a great river broken up into many channels by rocks and obstacles in its course.

The Milky Way is mainly made up of thousands

and thousands of small stars, and many more are revealed by the telescope; but, as we see in Cassiopeia, there are large bright stars in it too, though, of course, these may be infinitely nearer to us, and may only appear to us to be in the Milky Way because they are between us and it.

Now, besides the few constellations that I have mentioned, there are numbers of others, some of which are difficult to discover, as they contain no bright stars. But there are certain constellations which every one should know, because in them may be found some of the brightest stars, those of the first magnitude. Magnitude means size, and it is really absurd for us to say a star is of the first magnitude simply because it appears to us to be large, for, as I have explained already, a small star comparatively near to us might appear larger than a greater one further away. But the word 'magnitude' was used when men really thought stars were large or small according to their appearance, and so it is used to this day. They called the biggest and brightest first magnitude stars. Of these there are not many, only some twenty, in all the sky. The next brightest—about the brightness of the Pole Star and the stars in the Great Bear—are of the second magnitude, and so on, each magnitude containing stars less and less bright. When we come to stars of the sixth magnitude we have reached the limit of our sight, for seventh magnitude stars can only be seen with a telescope. Now that we understand what is meant by the magnitude, we can go back to the constellations and try to find some more.

If you draw an imaginary line across the two stars forming the backbone of the Bear, starting from the end nearest the tail, and continue it onward for a good distance, you will come to a very bright star called Capella, which you will know, because near it are three little ones in a triangle. Now, Capella means a goat, so the small ones are called the kids. In winter Capella gets high up into the sky, and then there is to be seen below her a little cluster called the Pleiades. There is nothing else like this in the whole sky. It is formed of six stars, as it appears to persons of ordinary sight, and these stars are of the sixth magnitude, the lowest that can be seen by the naked eye. But though small, they are set so close together, and appear so brilliant, twinkling like diamonds, that they are one of the most noticeable objects in the heavens. A legend tells that there were once seven stars in the Pleiades clearly visible, and that one has now disappeared. This is sometimes spoken of as 'the lost Pleiad,' but there does not seem to be any foundation for the story. In old days people attached particular holiness or luck to the number seven, and possibly, when they found that there were only six stars in this wonderful group, they invented the story about the seventh.

As the Pleiades rise, a beautiful reddish star of the first magnitude rises beneath them. It is called Aldebaran, and it, as well as the Pleiades, forms a part of the constellation of Taurus the bull. In England we can see in winter below Aldebaran the whole of the constellation of Orion, one of the finest of all the constellations, both for the number of the bright stars it contains and for the extent of the sky it covers. Four bright stars at wide distances enclose an irregular four-sided space in which are set three others close together and slanting downwards. Below these, again, are another three which seem to fall from them, but are not so bright. The figure of Orion as drawn in the old representations of the constellations is a very magnificent one. The three bright stars form his belt, and the three smaller ones the hilt of his sword hanging from it.

If you draw an imaginary line through the stars

ORION AND HIS NEIGHBOURS.

forming the belt and prolong it downwards slant-
ingly, you will see, in the very height of winter, the
brightest star in all the sky, either in the Northern
or Southern Hemisphere. This is Sirius, who stands
in a class quite by himself, for he is many times
brighter than any other first magnitude star. He
never rises very high above the horizon here, but
on crisp, frosty nights may be seen gleaming like ·
a big diamond between the leafless twigs and boughs
of the rime-encrusted trees. Sirius is the Dog Star,
and it is perhaps fortunate that, as he is placed, he
can be seen sometimes in the southern and sometimes
in the northern skies, so that many more people
have a chance of looking at his wonderful brilliancy,
than if he had been placed near the Pole star.
In speaking of the supreme brightness of Sirius
among the stars, we must remember that Venus
and Jupiter, which outrival him, are not stars, but
planets, and that they are much nearer to us.
Sirius is so distant that the measures for parallax
make hardly any impression on him, but, by repeated experiments, it has now been proved that
light takes more than eight years to travel from him
to us. So that, if you are eight years old, you are
looking at Sirius as he was when you were a baby!

Not far from the Pleiades, to the left as you face

them, are to be found two bright stars nearly the same size; these are the Heavenly Twins, or Gemini.

Returning now to the Great Bear, we find, if we draw a line through the middle and last stars of his tail, and carry it on for a little distance, we come fairly near to a cluster of stars in the form of a horse-shoe; there is only one fairly bright one in it, and some of the others are quite small, but yet the horse-shoe is distinct and very beautiful to look at. This is the Northern Crown. The very bright star not far from it is another first-class star called Arcturus.

To the left of the Northern Crown lies Hercules, which is only mentioned because near it is the point to which the sun with all his system appears at present to be speeding.

For other fascinating constellations, such as Leo or the Lion, Andromeda and Perseus, and the three bright stars by which we recognize Aquila the Eagle, you must wait awhile, unless you can get some one to point them out.

Those which you have noted already are enough to lead you on to search for more.

Perhaps some of you who live in towns and can see only a little strip of sky from the nursery or school-room windows have already found this chapter dull, and if so you may skip the rest of it and go on to the next. For the others, however, there is one more thing to know before leaving the subject, and that is the names of the string of constellations forming what is called the Zodiac. You may have heard the rhyme:

'The Ram, the Bull, the Heavenly Twins,
And next the Crab, the Lion shines,
The Virgin and the Scales;
The Scorpion, Archer, and He-goat,
The Man that holds the watering-pot,
The Fish with glittering tails.'

This puts in a form easy to remember the signs of the constellations which lie in the Zodiac, an imaginary belt across the whole heavens. It is very difficult to explain the Zodiac, but I must try. Imagine for a moment the earth moving round its orbit with the sun in the middle. Now, as the earth moves the sun will be seen continually against a different background—that is to say, he will appear to us to move not only across our sky in a day by reason of our rotation, but also along the sky, changing his position among the stars by reason of our revolution. You will say at once that we cannot see the stars when the sun is there, and no more we can. But the stars are there all the same, and every month the sun seems to have moved on into a new constellation, according to astronomers' reckoning.   If you count up the names of the constellations in the rhyme, you will find that there are just twelve, one for each month, and at the end of the year the sun has come round to the first one again.   The first one is Aries the Ram, and the sun is seen projected or thrown against that part of the sky where Aries is, in April, when we begin spring ; this is the first month to astronomers, and not January, as you might suppose.   Perhaps you will learn to recognize all the constellations in the Zodiac one day ; a few of them, such as the Bull and the Heavenly Twins, you know already if you have followed this chapter.

CHAPTER XII

WHAT THE STARS ARE MADE OF

How can we possibly tell what the stars are made of? If we think of the vast oceans of space lying between them and us, and realize that we can never cross those oceans, for in them there is no air, it would seem to be a hopeless task to find out anything about the stars at all. But even though we cannot traverse space ourselves, there is a messenger that can, a messenger that needs no air to sustain him, that moves more swiftly than our feeble minds can comprehend, and this messenger brings us tidings of the stars—his name is Light. Light tells us many marvellous things, and not the least marvellous is the news he gives us of the workings of another force, the force of gravitation. In some ways gravitation is perhaps more wonderful than light, for though light speeds across airless space, it is stopped at once by any opaque substance—that is to say, any substance not transparent, as you know very well by your own

159

shadows, which are caused by your bodies stopping the light of the sun.   Light striking on one side of the earth does not penetrate through to the other, whereas gravitation does.   You remember, of course, what the force of gravitation is, for we read about that very early in this book.   It is a mysterious attraction existing between all matter. Every atom pulls every other atom towards itself, more or less strongly according to distance.   Now, solid matter itself makes no difference to the force of gravitation, which acts through it as though it were not there.   The sun is pulling the earth toward itself, and it pulls the atoms on the far side of the earth just as strongly as it would if there were nothing lying between it and them. Therefore, unlike light, gravitation takes no heed of obstacles in the way, but acts in spite of them. The gravitation of the earth holds you down just the same, though you are on the upper floor of a house, with many layers of wood and plaster between you and it.   It cannot pull you down, for the floor holds you up, but it is gravitation that keeps your feet on the ground all the same.   A clever man made up a story about some one who invented a kind of stuff which stopped the force of gravitation going through it, just as a solid body stops light; when this stuff was made, of course, it went right away off into space, carrying with it anyone who stood on it, as there was nothing to hold it to the earth! That was only a story, and it is not likely anyone could invent such stuff, but it serves to make clear the working of gravitation. These two, light and gravitation, never tire, they run throughout the whole universe, and carry messages of tremendous importance for those who have minds to grasp them. Without light we could know nothing of these distant worlds, and without understanding the laws of gravity we should not be able to interpret much that light tells us.

To begin with light, what can we learn from it? We turn at once to our own great light-giver, the sun, to whom we owe not only all life, but also all the colour and beauty on earth. It is well known to men of science that colour lies in the light itself, and not in any particular object. That brilliant blue cloak of yours is not blue of itself, but because of the light that falls on it. If you cannot believe this, go into a room lighted only by gas, and hey, presto! the colour is changed as if it were a conjuring trick. You cannot tell now by looking at the cloak whether it is blue or green! Therefore

11

you must admit that as the colour changes with the change of light it must be due to light, and not to any quality belonging to the material of the cloak. But, you may protest, if the colour is solely due to light, and light falls on everything alike, why are there so many colours? That is a very fair question. If the light that comes from the sun were of only one colour—say blue or red—then everything would be blue or red all the world over. Some doors in houses are made with a strip of red or blue glass running down the sides. If you have one in your house like that, go and look through it, and you will see an astonishing world made up of different tones of the same colour. Everything is red or blue, according to the colour of the glass, and the only difference in the appearance of objects lies in the different shades, whether things are light or dark. This is a world as it might appear if the sun's rays were only blue or only red. But the sun's light is not of one colour only, fortunately for us; it is of all the colours mixed together, which, seen in a mass, make the effect of white light. Now, objects on earth are only either seen by the reflected light of the sun or by some artificial light. They have no light of their own. Put them in the dark and they do not shine at all; you cannot see them. It is the sun's light striking on them that makes them visible. But all objects do not reflect the light equally, and this is because they have the power of absorbing some of the rays that strike on them and not giving them back at all, and only those rays that are given back show to the eye. A white thing gives back all the rays, and so looks white, for we have the whole of the sun's light returned to us again. But how about a blue thing? It absorbs all the rays except the blue, so that the blue rays are the only ones that come back or rebound from it again to meet our eyes, and this makes us see the object blue; and this is the case with all the other colours. A red object retains all rays except the red, which it sends back to us; a yellow object gives back only the yellow rays, and so on. What an extraordinary and mysterious fact! Imagine a brilliant flower-garden in autumn. Here we have tall yellow sun-flowers with velvety brown centres, clustering pink and crimson hollyhocks, deep red and bright yellow peonies, slender fairy-like Japanese anemones, great bunches of mauve Michaelmas daisies, and countless others, and mingled with all these are many shades of green. Yet it is the light of the sun alone that falling on all these varied objects, makes that glorious blaze of colour ; it seems incredible. It may be difficult to believe, but it is true beyond all doubt. Each delicate velvety petal has some quality in it which causes it to absorb certain of the sun's rays and send back the others, and its colour is determined by those it sends back.

Well then how infinitely varied must be the colours hidden in the sun's light, colours which, mixed all together, make white light ! Yes, this is so, for *all* colours that we know are to be found there. In fact, the colours that make up sunlight are the colours to be seen in the rainbow, and they run in the same order. Have you ever looked carefully at a rainbow ? If not, do so at the next chance. You will see it begins by being dark blue at one end, and passes through all colours until it gets to red at the other.

We cannot see a rainbow every day just when we want to, but we can see miniature rainbows which contain just the same colours as the real ones in a number of things any time the sun shines. For instance, in the cut-glass edge of an inkstand or a decanter, or in one of those old-fashioned hanging pieces of cut-glass that dangle from the chandelier or candle-brackets. Of course you have often seen these colours reflected on the wall, and tried to get them to shine upon your face.   Or you have caught sight of a brilliant patch of colour on the wall and looked around to see what caused it, finally tracing it to some thick edge of shining glass standing in the sunlight.   Now, the cut-glass edge shows these colours to you because it breaks up the light that falls upon it into the colours it is made of, and lets each one come out separately, so that they form a band of bright colours instead of just one ray of white light.

This is perhaps a little difficult to understand, but I will try to explain.   When a ray of white light falls on such a piece of glass, which is known as a prism, it goes in as white light at one side, but the three-cornered shape of the glass breaks it up into the colours it is made of, and each colour comes out separately at the other side—namely, from blue to red—like a little rainbow, and instead of one ray of white light, we have a broad band of all the colours that light is made of.

Who would ever have thought a pretty plaything like this could have told us what we so much wanted to know—namely, what the sun and the stars are made of?   It seems too marvellous to be true, yet true it is that for ages and ages light has been carrying its silent messages to our eyes, and only recently men have learnt to interpret them. It is as if some telegraph operator had been going steadily on, click, click, click, for years and years, and no one had noticed him until someone learnt the code of dot and dash in which he worked, and then all at once what he was saying became clear. The chief instrument in translating the message that the light brings is simply a prism, a three-cornered wedge of glass, just the same as those hanging lustres belonging to the chandeliers. When a piece of glass like this is fixed in a telescope in such a way that the sun's rays fall on it, then there is thrown on to a piece of paper or any other suitable background a broad coloured band of lovely light like a little rainbow, and this is called the sun's spectrum, and the instrument by which it is seen is called a spectroscope. But this in itself could tell us little ; the message it brings lies in the fact that when it has passed through the telescope, so that it is magnified, it is crossed by hundreds of minute black lines, not placed evenly at all, but scattered up and down. There may be two so close together that they look like one, and then three far apart, and then some more at different distances. When this remarkable appearance was examined carefully it was found that in sunlight the lines that appeared were always exactly the same, in the same places, and this seemed so curious that men began to seek for an explanation.

Someone thought of an experiment which might teach us something about the matter, and instead of letting sunlight fall on the prism, he made an artificial light by burning some stuff called sodium, and then allowed the band of coloured light to pass through the telescope ; when he examined the spectrum that resulted, he found that, though numbers of lines to be found in the sun's spectrum were missing, there were a few lines here exactly matching a few of the lines in the sun's spectrum ; and this could not be the result of chance only, for the lines are so mathematically exact, and are in themselves so peculiarly distributed, that it could only mean that they were due to the same cause. What could this signify, then, but that away up there in the sun, among other things, stuff called sodium, very well known to chemists on earth, is burning ? After this many other substances were heated white-hot so as to give out light, in order to discover if the lines to be seen in their spectra were also to be found in the sun's spectrum. One of these was iron, and, astonishing to say, all the many little thread-like lines that appeared in its spectrum were reproduced to a hair's-breadth, among others, in the sun's spectrum.

So we have found out beyond all possibility of doubt some of the materials of which the sun is made. We know that iron, sodium, hydrogen, and numerous other substances and elements, are all burning away there in a terrific furnace, to which any furnace we have on earth is but as the flicker of a match.

It was not, of course, much use applying this method to the planets, for we know that the light which comes from them to us is only reflected sunlight, and this, indeed, was proved by means of the spectroscope. But the stars shine by their own light, and this opened up a wide field for inquiry. The difficulty was, of course, to get the light of one star separated from all the rest, because the light of one star is very faint and feeble to cast a spectrum at all. Yet by infinite patience difficulties were overcome. One star alone was allowed to throw its light into the telescope ; the light passed through a prism, and showed a faint band of many colours, with the expected little black lines cutting across it more or less thickly. Examinations have thus been made of hundreds of stars. In the course of them some substances as yet unknown to us on earth have been encountered, and in some stars one element—hydrogen—is much stronger than in others ; but, on the whole, speaking broadly, it has been satisfactorily shown that the stars are made

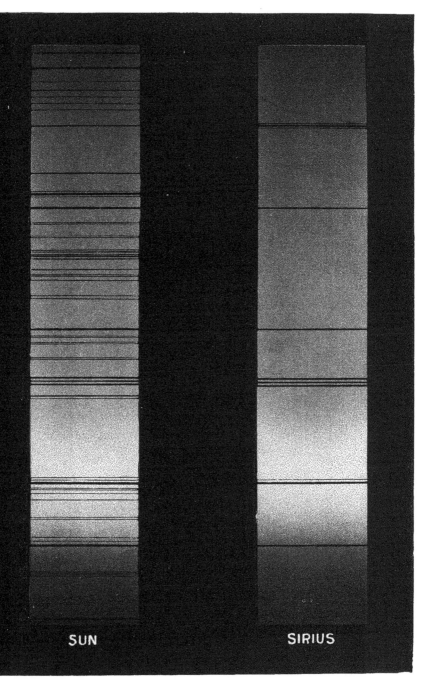

SUN   SIRIUS

on the same principles as our own sun, so that the reasoning of astronomers which had argued them to be suns was proved.

We have here in the picture the spectrum of the sun and the spectrum of Sirius. You can see that the lines which appear in the band of light belonging to Sirius are also in the band of light belonging to the sun, together with many others. This means that the substances flaming out and sending us light from the far away star are also giving out light from our own sun, and that the sun and Sirius both contain the same elements in their compositions.

This, indeed, seems enough for the spectroscope to have accomplished; it has interpreted for us the message light brings from the stars, so that we know beyond all possibility of mistake that these glowing, twinkling points of light are brilliant suns in a state of intense heat, and that in them are burning elements with which we ourselves are quite familiar. But when the spectroscope had done that, its work was not finished, for it has not only told us what the stars are made of, but another thing which we could never have known without it—namely, if they are moving toward us or going away from us.

CHAPTER XIII

RESTLESS STARS

You remember we have already remarked upon the difficulty of telling how far one star lies behind another, as we do not know their sizes. It is, to take another similar case, easy enough to tell if a star moves to one side or the other, but very difficult by ordinary observation to tell if it is advancing toward us or running away from us, for the only means we have of judging is if it gets larger or smaller, and at that enormous distance the fact whether it advances or recedes makes no difference in its size. Now, the spectroscope has changed all this, and we can tell quite as certainly if a star is coming toward us as we can if it moves to one side. I will try to explain this. You know, perhaps, that sound is caused by vibration in the air. The noise, whatever it is, jars the air and the vibrations strike on our ears. It is rather the same thing as the result of throwing a stone into a pond: from the centre of the splash little wavelets run out in ever-widening circles; so through the air run ever-widening vibrations from every sound. The more vibrations there are in a second the shriller is the note they make. In a high note the air-vibrations follow one another fast, pouring into one's ear at a terrific speed, so that the apparatus in the ear which receives them itself vibrates fiercely and records a high note, while a lower note brings fewer and slower vibrations in a second, and the ear is not so much disturbed. Have you ever noticed that if a railway engine is sweeping toward you and screaming all the time, its note seems to get shriller and shriller? That is because the engine, in advancing, sends the vibrations out nearer to you, so more of them come in a second, and thus they are crowded up closer together, and are higher and higher.

Now, light is also caused by waves, but they are not the same as sound waves. Light travels without air, whereas sound we know cannot travel without air, and is ever so much slower, and altogether a grosser, clumsier thing than light. But yet the waves or rays which make light correspond in some ways to the vibrations of sound. What corresponds to the treble on the piano is the blue end of the spectrum in light, and the bass is the red end. Now, when we are looking at the spectrum of any body which is advancing swiftly toward us, something of the same effect is observed as in the case of the shrieking engine.   Take any star and imagine that that star is hastening toward us at a pace of three hundred miles a second, which is not at all an unusual rate for a star; then, if we examine the band of light, the spectrum, of such a star, we shall observe an extraordinary fact—all these little lines we have spoken of are shoved up toward the treble or blue end of the spectrum.   They still remain just the same distances from each other, and are in twos and threes or single, so that the whole set of lines is unaltered as a set, but every one of them is shifted a tiny fraction up toward the blue end of the spectrum, just a little displaced.   Now if, instead of advancing toward us, this same star had been rushing away from us at a similar pace, all these lines would have been moved a tiny bit toward the red or bass end of the spectrum.   This is known to be certainly true, so that by means of the spectroscope we can tell that some of these great sun-stars are advancing toward us and some receding from us, according to whether the multitudes of little lines in the spectrum are shifted slightly to the blue or the red end.

You remember that it has been surmised that the pace the sun moves with his system is about twelve miles a second. This seems fast enough to us, who think that one mile a minute is good time for an express train, but it is slow compared with the pace of many of the stars. As I have said, some are travelling at a rate of between two hundred and three hundred miles a second; and it is due to the spectroscope that we know not only whether a star is advancing toward us or receding from us, but also whether the pace is great or not; it even tells us what the pace is, up to about half a mile a second, which is very marvellous. It is a curious fact that many of the small stars show greater movement than the large ones, which may or may not mean that they are nearer to us.

It may be taken as established that there is no such thing as absolute rest in the universe: everything, stars and nebulæ alike, are moving somewhere; in an infinite variety of directions, with an infinite variety of speed they hasten this way and that. It would be impossible for any to remain still, for even supposing it had been so 'in the beginning,' the vast forces at work in the universe would not let it remain so. Out of space would come the persistent call of gravitation: atoms would cry silently to atoms. There could be no perfect equality of pull on all sides; from one side or another the pull would be the stronger. Slowly the inert mass would obey and begin falling toward it; it might be an inch at a time, but with rapid increase, until at last it also was hastening some whither in this universe which appears to us to be infinite.

It must be remembered that these stars, even when moving at an enormous pace, do not change their places in the sky when regarded by ordinary observers. It would take thousands of years for any of the constellations to appear at all different from what they are now, even though the stars that compose them are moving in different directions with a great velocity, for a space of many millions of miles, at the distance of most of the stars, would be but as the breadth of a fine hair as seen by us on earth. So thousands of years ago men looked up at the Great Bear, and saw it apparently the same as we see it now; yet for all that length of time the stars composing it have been rushing in this direction and that at an enormous speed, but do not appear to us on the earth to alter their positions in regard to each other. I know of nothing that gives one a more overwhelming sense of the mightiness of the universe and the smallness of ourselves than this fact. From age to age men look on changeless heavens, yet this apparently stable universe is fuller of flux and reflux than is the restless ocean itself, and the very wavelets on the sea are not more numerous nor more restless than the stars that bestrew the sky.

CHAPTER XIV

THE COLOURS OF THE STARS

Has it ever occurred to you that the stars are not all of the same colour ? It is true that, just glancing at them casually, you might say they are all white ; but if you examine them more carefully you cannot help seeing that some shine with a steely blue light, while others are reddish or yellowish. These colours are not easy to distinguish with the naked eye, and might not attract any attention at all unless they were pointed out ; yet when attention is drawn to the fact, it is impossible to deny the redness of some, such as Aldebaran. But though we may admit this, we might add that the colours are so very faint and inconspicuous, that they might be, after all, only the result of imagination.

To prove that the colours are constant and real we must use a telescope, and then we need have no further doubt of their reality, for instead of disappearing, the colours of some stars stand out

176

quite vividly beyond the possibility of mistake. Red stars are a bright red, and they are the most easily seen of all, though the other colours, blue and yellow and green, are seen very decidedly by some people. The red stars have been described by various observers as resembling ' a drop of blood on a black field,' ' most magnificent copper-red,' ' most intense blood-red,' and ' glowing like a live coal out of the darkness of space.' Some people see them as a shining red, like that of a glowing cloud at sunset. Therefore there can be no doubt that the colours are genuine enough, and are telling us some message. This message we are able to read, for we have begun to understand the language the stars speak to us by their light since the invention of the spectroscope. The spectroscope tells us that these colours indicate different stages in the development of the stars, or differences of constitution—that is to say, in the elements of which they are made. Our own sun is a yellow star, and other yellow stars are akin to him ; while red and blue and green stars contain different elements, or elements m different proportions.

Stars do not always remain the same colours for an indefinite time ; one star may change slowly

12

from yellow to white, and another from red to yellow; and there are instances of notable changes, such as that of the brilliant white Sirius, who was stated in old times by many different observers to be a red star. All this makes us think, and year by year thought leads us on to knowledge, and knowledge about these distant suns increases. But though we know a good deal now, there are still many questions we should like to ask which we cannot expect to have answered for a long time yet, if ever.

The star colours have some meanings which we cannot even guess; we can only notice the facts regarding them. For instance, blue stars are never known to be solitary—they always have a companion, but why this should be so passes our comprehension. What is it in the constitution of a blue star which holds or attracts another? Whatever it may be, it is established by repeated instances that blue stars do not stand alone. In the constellation of Cygnus there are two stars, a blue and a yellow one, which are near enough to each other to be seen in the same telescope at the same time, and yet in reality are separated by an almost incredible number of billions of miles. But as we know that a blue star is never seen alone, and that it has often as its companion a yellowish or reddish star, it is probable that these two, situated at an enormous distance from one another, are yet in some mysterious way dependent on each other, and are not merely seen together because they happen to fall in the same field of view.

Many double stars show most beautifully contrasted colours : among them are pairs of yellow and rose-red, golden and azure, orange and purple, orange and lilac, copper-colour and blue, apple-green and cherry-red, and so on. In the Southern Hemisphere there is a cluster containing so many stars of brilliant colours that Sir John ·Herschel named it ' the Jewelled Cluster.'

I expect most of you have seen an advertisement of Pear's Soap, in which you are asked to stare at some red letters, and then look away to some white surface, such as a ceiling, when you will see the same letters in green. This is because green is the complementary or contrasting colour to red, and the same thing is the case with blue and yellow. When any one colour of either of these pairs is seen, it tends to make the other appear by reaction, and if the eye gazed hard at blue instead of red, it would next see yellow, and not green. Now, many people to whom this curious fact is known argue that perhaps the colours of the double stars are not real, but the effect of contrast only ; for instance, they say a red star near a companion white one would tend to make the companion appear green, and so, of course, it would.   But this does not account for the star colours, which are really inherent in the stars themselves, as may be proved by cutting off the light of one star, and looking only at the other, when its colour still appears unchanged.   Another argument equally strong against the contrast theory is that the colours of stars in pairs are by no means always those which would appear if the effect was only due to complementary colours.   It is not always blue and yellow or red and green pairs that we see, though these are frequent, but many others of various kinds, such as copper and blue, and ruddy and blue.

We have therefore come to the conclusion that there are in this astonishing universe numbers of gloriously coloured suns, some of which apparently lie close together.   What follows ?   Why, we want to know, of course, if these stars are really pairs connected with each other, or if they only appear so by being in the same line of sight, though one is infinitely more distant than the other.   And that question also has been answered.   There are now known thousands of cases in which stars, hitherto regarded as single, have been separated into two, or even more, by the use of a telescope. Of these thousands, some hundreds have been carefully investigated, and the result is that, though there are undoubtedly some in which the connexion is merely accidental, yet in by far the greater number of cases the two stars thus seen together have really some connexion which binds them to one another ; they are dependent on one another. This has been made known to us by the working of the wonderful law of gravitation, which is obeyed throughout the whole universe. We know that by the operation of this law two mighty suns will be drawn toward each other with a certain pull, just as surely as we know that a stone let loose from the hand will fall upon the earth ; so by noting the effect of two mighty suns upon each other many facts about them may be found out. By the most minute and careful measurements, by the use of the spectroscope, and by every resource known to science, astronomers have, indeed, actually found out with a near approach to exactness how far some of these great suns lie from each other, and how large they are in comparison with one another.

The very first double star ever discovered was

one which you have already seen, the middle one in the tail of the Great Bear. If you look at it you will be delighted to find that you can see a wee star close to it, and you will think you are looking at an example of a double star with your very own eyes; but you will be wrong, for that wee star is separated by untold distances from the large one to which it seems so near. In fact, any stars which can be seen to be separate by the naked eye must lie immeasurably far apart, however tiny seems the space between them. Such stars may possibly have some connexion with each other, but, at any rate in this case, such a connexion has not been proved. No, the larger star itself is made up of two others, which can only be seen apart in a telescope. Since this discovery double stars have been plentifully found in every part of the sky. The average space between such double stars as seen from our earth is—what do you think? It is the width of a single hair held up thirty-six feet from our eyes! This could not, of course, be seen without the use of a telescope or opera-glasses. It serves to give some impression of star distances when we think that the millions and millions of miles lying between those stars have shrunk to that hair's-breadth seen from our point of view.

Twin stars circle together round a common centre of gravity, and are bound by the laws of gravitation just as the planets are. Our sun is a solitary star, with no companion, and therefore such a state of things seems to us to be incredible. Fancy two gigantic suns, one topaz-yellow and the other azure-blue, circling around in endless movement! Where in such a system would there be room for the planets? How could planets exist under the pull of two suns in opposite directions? Still more wonders are unfolded as the inquiry proceeds. Certain irregularities in the motions of some of these twin systems led astronomers to infer that they were acted upon by another body, though this other body was not discernible. In fact, though they could not see it, they knew it must be there, just as Adams and Leverrier knew of the existence of Neptune, before ever they had seen him, by the irregularities in the movements of Uranus. As the results showed, it was there, and was comparable in size to the twin suns it influenced, and yet they could not see it. So they concluded this third body must be dark, not light-giving like its companions. We are thus led to the strange conclusion that some of these systems are very complicated, and are formed not only of shining suns, but of huge dark bodies which cannot be called suns. What are they, then? Can they be immense planets? Is it possible that life may there exist? No fairy tale could stir the imagination so powerfully as the thought of such systems including a planetary body as large or larger than its sun or suns. If indeed life exists there, what a varied scene must be presented day by day! At one time both suns mingling their flashing rays may be together in the sky; at another time only one appears, a yellow or blue sun, as the case may be. The surface of such planets must undergo weird transformations, the foliage showing one day green, the next yellow, and the next blue; shadows of azure and orange will alternate! But fascinating as such thoughts are, we can get no further along that path.

To turn from fancy to facts, we find that telescope and spectroscope have supplied us with quite enough matter for wonder without calling upon imagination. We have discovered that many of the stars which seem to shine with a pure single light are double, and many more consist not only of two stars, but of several, some of which may be dark bodies. The Pole Star was long known to be double, and is now discovered to have a third member in its system. These multiple systems vary from one another in almost every case. Some are made up of a mighty star and a comparatively small one; others are composed of stars equal in light-giving power—twin suns. Some progress swiftly round their orbits, some go slowly; indeed, so slowly that during the century they have been under observation only the very faintest sign of movement has been detected; and in other systems, which we are bound to suppose double, the stars are so slow in their movements that no progress seems to have been made at all.

The star we know as the nearest to us in the heavens, Alpha Centauri, is composed of two very bright partners, which take about eighty-seven years to traverse their orbit. They sometimes come as near to each other as Saturn is to the sun. In the case of Sirius astronomers found out that he had a companion by reason of his irregularities of movement before they discovered that companion, which is apparently a very small star, only to be discerned with good telescopes. But here, again, it would be unwise to judge only by what we see. Though the star appears small, we know by the influence it exercises on Sirius that it is very nearly the same size as he is. Thus we judge that it is poor in light-giving property ; in fact, its shining power is much less than that of its companion, though its size is so nearly equal.   This is not wonderful, for Sirius's marvellous light-giving power is one of the wonders of the universe ; he shines as brilliantly as twenty-nine or thirty of our suns !

In some cases the dark body which we cannot see may even be larger than the shining one, through which alone we can know anything of it.   Here we have a new idea, a hint that in some of these systems there may be a mighty earth with a smaller sun going round it, as men imagined our sun went around the earth before the real truth was found out.

So we see that, when we speak of the stars as suns comparable with our sun, we cannot think of them all as being exactly on the same model. There are endless varieties in the systems ; there are solitary suns like ours which may have a number of small planets going round them, as in the solar system; but there are also double suns going round each other, suns with mighty dark bodies revolving round them which may be planets, and huge dark bodies with small suns too.   Every increase of knowledge opens up new wonders, and the world in which we live is but one kind of world amid an infinite number.

In this chapter we have learnt an altogether new fact—the fact that the hosts of heaven comprise not only those shining stars we are accustomed to see, but also dark bodies equally massive, and probably equally numerous, which we cannot see. In fact, the regions of space may be strewn with such dark bodies, and we could have no possible means of discovering them unless they were near enough to some shining body to exert an influence upon it. It is not with his eyes alone, or with his senses, man knows of the existence of these great worlds, but often solely by the use of the powers of his mind.

CHAPTER XV

TEMPORARY AND VARIABLE STARS

It is a clear night, nearly all the world is asleep, when an astronomer crosses his lawn on his way to his observatory to spend the dark hours in making investigations into profound space. His brilliant mind, following the rays of light which shoot from the furthest star, will traverse immeasurable distances, while the body is forgotten. Just before entering the observatory he pauses and looks up; his eye catches sight of something that arrests him, and he stops involuntarily. Yet any stranger standing beside him, and gazing where he gazes, would see nothing unusual. There is no fiery comet with its tail stretching across from zenith to horizon, no flaming meteor dashing across the darkened sky. But that there is something unusual to be seen is evident, for the astronomer breathes quickly, and after another earnest scrutiny of the object which has attracted him, he rushes into the observatory, searches for a star-chart, and

examines attentively that part of the sky at which
he has been gazing. He runs his finger over the
chart: here and there are the well-known stars that
mark that constellation, but here? In that part
there is no star marked, yet he knows, for his own
eyes have told him but a few moments ago, that
here there is actually blazing a star, not large,
perhaps, but clear enough to be seen without a
telescope—a star, maybe, which no eye but his has
yet observed!

He hurries to his telescope, and adjusts it so as
to bring the stranger into the field of view. A
new star! Whence has it come? What does it
mean?

By the next day at the latest the news has flown
over the wires, and all the scientific world is aware
that a new star has been detected where no star
ever was seen before. Hundreds of telescopes are
turned on to it; its spectrum is noted, and it
stands revealed as being in a state of conflagration,
having blazed up from obscurity to conspicuousness.
Night after night its brilliance grows, until it ranks
with the brightest stars in heaven, and then it dies
down and grows dim, gradually sinking—sinking into
the obscurity from whence it emerged so briefly, and
its place in the sky knows it no more. It may be

there still, but so infinitely faint and far away that no power at our command can reveal it to us. And the amazing part of it is that this huge disaster, this mighty conflagration, is not actually happening as it is seen, but has happened many hundreds of years ago, though the message brought by the light carrier has but reached us now.

There have not been a great many such outbursts recorded, though many may have taken place unrecorded, for even in these days, when trained observers are ceaselessly watching the sky, ' new' stars are not always noticed at once. In 1892 a new star appeared, and shone for two months before anyone noticed it. This particular one never rose to any very brilliant size. It was situated in the constellation of Auriga, and was noticed on February 1. It remained fairly bright until March 6, when it began to die down; but it has now sunk so low that it can only be seen in the very largest telescopes.

Photography has been most useful in recording these stars, for when one is noticed it has sometimes been found that it has been recorded on a photographic plate taken some time previously, and this shows us how long it has been visible. More and more photography becomes the useful handmaid of astronomers, for the photographic prepared plate is more sensitive to rays of light than the human eye, and, what is more useful still, such plates retain the rays that fall upon them, and fix the impression. Also on a plate these rays are cumulative—that is to say, if a very faint star shines continuously on a plate, the longer the plate is exposed, within certain limits, the clearer will the image of that star become, for the light rays fall one on the top of the other, and tend to enforce each other, and so emphasize the impression, whereas with our eyes it is not the same thing at all, for if we do not see an object clearly because it is too faint, we do not see it any better, however much we may stare at the place where it ought to be. This is because each light ray that reaches our eye makes its own impression, and passes on; they do not become heaped on each other, as they do on a photographic plate.

One variable star in Perseus, discovered in 1901, rose to such brilliancy that for one night it was queen of the Northern Hemisphere, outshining all the other first-class stars.

It rose into prominence with wonderful quickness, and sank equally fast. At its height it outshone our sun eight thousand times! This star was so far from us that it was reckoned its light must take about three hundred years to reach us, consequently the great conflagration, or whatever caused the outburst, must have taken place in the reign of James I., though, as it was only seen here in 1901, it was called the new star of the new century.

When these new stars die down they sometimes continue to shine faintly for a long time, so that they are visible with a telescope, but in other cases they may die out altogether. We know very little about them, and have but small opportunity for observing them, and so it is not safe to hazard any theories to account for their peculiarities. At first men supposed that the great flame was made by a violent collision between two bodies coming together with great velocity so that both flared up, but this speculation has been shown by the spectroscope to be improbable, and now it is supposed by some people that two stars journeying through space may pass through a nebulous region, and thus may flare up, and such a theory is backed up by the fact that a very great number of such stars do seem to be mixed up in some strange way with a nebulous haze.

All these new stars that we have been discussing

so far have only blazed up once and then died down, but there is another class of stars quite as peculiar, and even more difficult to explain, and these are called variable stars.   They get brighter and brighter up to a certain point, and then die down, only to become bright once more, and these changes occur with the utmost regularity, so that they are known and can be predicted beforehand. This is even more unaccountable than a sudden and unrepeated outburst, for one can understand a great flare-up, but that a star should flare and die down with regularity is almost beyond comprehension.   Clearly we must look further than before for an explanation.   Let us first examine the facts we know.   Variable stars differ greatly from each other.   Some are generally of a low magnitude, and only become bright for a short time, while others are bright most of the time and die down only for a short time.   Others become very bright, then sink a little bit, but not so low as at first; then they become bright again, and, lastly, go right down to the lowest point, and they keep on always through this regular cycle of changes.   Some go through the whole of these changes in three days, and others take much longer.   The periods, as the intervals between

13

the complete round of changes are called, vary, in fact, between three days and six hundred! It may seem impossible that changes covering so long as six hundred days could be known and followed, but there is nothing that the patience of astronomers will not compass.

One very well-known variable star you can see for yourselves, and as an ounce of observation is worth a pound of hearsay, you might take a little trouble to find it. Go out on any clear starlight night and look. Not very far from Cassiopeia (W.), to the left as you face it, are three bright stars running down in a great curve. These are in the constellation called Perseus, and a little to the right of the middle and lowest one is the only variable star we can see in the sky without a telescope.

This is Algol. For the greater part of three days he is a bright star of about the second magnitude, then he begins to fade, and for four and a half hours grows steadily dimmer. At the dimmest he remains for about twenty minutes, and then rises again to his ordinary brightness in three and a half hours. How can we explain this? You may possibly be able to suggest a reason. What do you say to a dark body revolving round Algol, or, rather, revolving with him round a common centre of gravity ?   If such a thing were indeed true, and if such a body happened to pass between us and Algol at each revolution, the light of Algol would be cut off or eclipsed in proportion to the size of such a body.   If the dark body were the full size of Algol and passed right between him and us, it would cut off all the light, but if it were not quite the same size, a little would still be seen. And this is really the explanation of the strange changes in the brightness of Algol, for such a dark body as we are imagining does in reality exist.   It is a large dark body, very nearly as large as Algol himself, and if, as we may conjecture, it is a mighty planet, we have the extraordinary example of a planet and its sun being nearly the same size.   We have seen that the eclipse happens every three days, and this means, of course, that the planetary body must go round its sun in that time, so as to return again to its position between us and him, but the thing is difficult to believe.   Why, the nearest of all our planets to the sun, the wee Mercury, takes eighty-seven days to complete its orbit, and here is a mighty body hastening round its sun in three !   To do this in the time the large dark planet must be very near to Algol ; indeed, astronomers have calculated that the surfaces of the two bodies are not more than about two million miles apart, and this is a trifle when we consider that we ourselves are more than forty-six times as far as that from the sun.   At this distance Algol, as observed from the planet, will fill half the sky, and the heat he gives out must be something stupendous.   Also the effects of gravitation must be queer indeed, acting on two such huge bodies so close together.   If any beings live in such a strange world, the pull which draws them to their mighty sun must be very nearly equal to the pull which holds them to their own globe; the two together may counteract each other, but the effect must be strange!

From irregularities in the movements of Algol it has been judged that there may be also in the same system another dark body, but of it nothing has been definitely ascertained.

But all variable stars need not necessarily be due to the light being intercepted by a dark body. There are cases where two bright stars in revolving round each other produce the same effect; for when seen side by side the two stars give twice as much light as when one is hidden behind the other, and as they are seen alternately side by side and in line, they seem to alter regularly in lustre.

CHAPTER XVI

STAR CLUSTERS AND NEBULÆ

COULD you point out any star cluster in the sky? You could if you would only think for a minute, for one has been mentioned already. This is the cluster known as the Pleiades, and it is so peculiar and so different from anything else, that many people recognize the group and know where to look for it even before they know the Great Bear, the favourite constellation in the northern sky, itself. The Pleiades is a real star cluster, and the chief stars in it are at such enormous distances from one another that they can be seen separately by the eye unaided, whereas in most clusters the stars appear to be so close together that without a telescope they make a mere blur of brightness. For a long time it was supposed that the stars composing the Pleiades could not really be connected because of the great distances between them; for, as you know, even a hair's-breadth apparently between stars signifies in reality many millions of miles.

Light travelling from the Pleiades to us, at that incomprehensible pace of which you already know, takes a hundred and ninety years to reach us! At this incredibly remote distance lies the main part of the cluster from us ; but it is more marvellous still that we have every reason to believe that the outlying stars of this cluster are as far from the central ones as the nearest star we know of, Alpha Centauri, is from us! Little wonder was it, then, that men hesitated to ascribe to the Pleiades any real connection with each other, and supposed them to be merely an assemblage of stars which seemed to us to lie together.

With the unaided eye we see comparatively few stars in the Pleiades. Six is the usual number to be counted, though people with very good sight have made out fourteen. Viewed through the telescope, however, the scene changes: into this part of space stars are crowded in astonishing profusion ; it is impossible to count them, and with every increase in the power of the telescope still more are revealed. Well over a thousand in this small space seems no exaggerated estimate. Now, it is impossible to say how many of these really belong to the group, and how many are seen there accidentally, but observations of the most prominent ones have shown that they are all moving in exactly the same direction at the same pace. It would be against probability to conceive that such a thing could be the result of mere chance, considering the infinite variety of star movements in general, and so we are bound to believe that this wonderful collection of stars is a real group, and not only an apparent one.

So splendid are the great suns that illuminate this mighty system, that at least fifty or sixty of them far surpass our own sun in brilliancy. Therefore when we look at that tiny sparkling group we must in imagination picture it as a vast cluster of mighty stars, all controlled and swayed by some dominant impulse, though separated by spaces enough to make the brain reel in thinking of them. If these suns possess also attendant planets, what a galaxy of worlds, what a universe within a universe is here!

Other star clusters there are, not so conspicuous as the Pleiades, and most of these can only be seen through a telescope, so we may be thankful that we have one example so splendid within our own vision. There are some clusters so far and faintly shining that they were at first thought to be nebulæ, and not stars at all; but the telescope gradually revealed the fact that many of these are made up

of stars, and so people began to think that all faint
shining patches of nebulous light were really star
clusters, which would be resolved into stars if only
we had better telescopes. Since the invention of
the spectroscope, however, fresh light has been
thrown on the matter, for the spectrum which is
shown by some of the nebulous patches is not the
same as that shown by stars, and we know that
many of these strange appearances are not made
up of infinitely distant stars.

We are talking here quite freely about nebulæ
because we have met one long ago when we discussed
the gradual evolution of our own system, and we
know quite well that a nebula is composed of
luminous faintly-glowing gas of extreme fineness
and thinness. We see in the sky at the present
time what we may take to be object-lessons in our
own history, for we see nebulæ of all sorts and
sizes, and in some stars are mixed up, and in others
stars are but dimly seen, so that it does not require
a great stretch of the imagination to picture these
stars as being born, emerging from the swaddling
bands of filmy webs that have enwrapped them ; and
other nebulæ seem to be gas only, thin and glowing, with no stars at all to be found in it. We still
know very little about these mysterious appearances,

but the work of classifying and resolving them is going on apace. Nebulæ are divided into several classes, but the easiest distinction to remember is that between white nebulæ and green nebulæ. This is not to say that we can see some coloured green, but that green appears in the spectrum of some of the nebulæ, while the spectrum of a white nebula is more like that of a star.

It is fortunate for us that in the sky we can see without a telescope one instance of each of the several objects of interest that we have referred to.

We have been able to see one very vivid example of a variable star ; we have seen one very beautiful example of a star cluster ; and it remains to look for one very good example of a white nebula.

Just as in finding Algol you were doing a little bit of practical work, proving something of which you had read, so by seeing this nebula you will remember more about nebulæ in general than by reading many chapters on the subject. This particular nebula is in Andromeda, and is not far from Algol ; and it is not difficult to find. It is the only one that can be well seen without a telescope, and was known to the ancients ; it is believed to have been mentioned in a book of the tenth century !

If you take an imaginary line down from the two left-hand stars of Cassiopeia, and follow it carefully, you will come before long to a rather faint star, and close to it is the nebula.

When you catch sight of it you will, perhaps, at first be disappointed, for all you will see is a soft blur of white, as if someone had laid a dab of luminous paint on the sky with a finger; but as you gaze at it night after night and realize its unchangeableness, realize also that it is a mass of glowing gas, an island in space, infinitely distant, unsupported and inexplicable, something of the wonder of it will creep over you.

Thousands of telescopic nebulæ are now known, and have been examined, and they are of all shapes. Roughly, they have been divided up into several classes—those that seem to us to be round and those that are long ovals, like this one in Andromeda; but these may, of course, be only round ones seen edgewise by us; others are very irregular, and spread over an enormous part of the sky. The most remarkable of these is that in Orion, and if you look very hard at the middle star in the sword-hilt of Orion, you may be able to make out a faint mistiness. This, when seen through a telescope, becomes a wonderful and far-spreading

THE GREAT NEBULA IN ANDROMEDA.

nebula, with brighter and darker parts like gulfs in it, and dark channels. It has been sometimes called the Fish-mouth Nebula, from a fanciful idea as to its shape. Indeed, so extraordinarily varied are these curious structures, that they have been compared with numbers of different objects. We have some like brushes, others resembling fans, rings, spindles, keyholes ; others like animals—a fish, a crab, an owl, and so on ; but these suggestions are imaginative, and have nothing to do with the real problem. In *The System of the Stars* Miss Clerke says: 'In regarding these singular structures we seem to see surges and spray-flakes of a nebulous ocean, bewitched into sudden immobility; or a rack of tempest-driven clouds hanging in the sky, momentarily awaiting the transforming violence of a fresh onset. Sometimes continents of pale light are separated by narrow straits of comparative darkness ; elsewhere obscure spaces are hemmed in by luminous inlets and channels.'

One curious point about the Orion Nebula is that the star which seems to be in the midst of it resolves itself under the telescope into not one but six, of various sizes.

Nebulæ are in most cases too enormously remote from the earth for us to have any possible means of computing the distance ; but we may take it that light must journey at least a thousand years to reach us from them, and in many cases much more. Therefore, if at the time of the Norman Conquest a nebula had begun to grow dim and fade away, it would, for all intents and purposes, still be there for us, and for those that come after us for several generations, though all that existed of it in reality would be its pale image fleeting onward through space in all directions in ever-widening circles.

That nebulæ do sometimes change we have evidence: there . are cases in which some have grown indisputably brighter during the years they have been under observation, and some nebulæ that have been recorded by careful observers seem to have vanished. When we consider that these strange bodies fill many, many times the area of our whole solar system to the outermost bounds of Neptune's orbit, it is difficult to imagine what force it is that acts on them to revive or quench their light. That that light is not the direct result of heat has long been known ; it is probably some form of electric excitement causing luminosity, very much as it is caused in the comets. Indeed, many people have been tempted to think of the nebulæ as the comets of the universe, and in some points there are, no doubt, strong resemblances between the two. Both shine in the same way, both are so faint and thin that stars can be seen through them; but the spectroscope shows us that to carry the idea too far would be wrong, as there are many differences in constitution.

We have seen that there are dark stars as well as light stars; if so, may there not be dark nebulæ as well as light ones? It may very well be so. We have seen that there are reasons for supposing our own system to have been at first a cool dark nebula rotating slowly. The heavens may be full of such bodies, but we could not discern them. Their thinness would prevent their hiding any stars that happened to be behind them. No evidence of their existence could possibly be brought to us by any channel that we know.

It is true that, besides the dark rifts in the bright nebulæ, which may themselves be caused by a darker and non-luminous gas, there are also strange rifts in the Milky Way, which at one time were conjectured to be due to a dark body intervening between us and the starry background. This idea is now quite discarded; whatever may cause them, it is not that. One of the most startling of these rifts is that called the Coal-Sack, in the Southern Hemisphere, and it occurs in a part of the sky otherwise so bright that it is the more noticeable. No possible explanation has yet been suggested to account for it.

Thus it may be seen that, though much has been discovered, much remains to be discovered. By the patient work of generations of astronomers we have gained a clear idea of our own position in the universe. Here are we on a small globe, swinging round a far mightier and a self-luminous globe, in company with seven other planets, many of which, including ourselves, are attended by satellites or moons. Between the orbits of these planets is a ring or zone of tiny bodies, also going round the sun. Into this system flash every now and then strange luminous bodies—some coming but once, never to return ; others returning again and again.

Far out in space lies this island of a system, and beyond the gulfs of space are other suns, with other systems : some may be akin to ours and some quite different. Strewn about at infinite distances are star clusters, nebulæ, and other mysterious objects.

The whole implies design, creation, and the

working of a mighty intelligence ; and yet there are
small, weak creatures here on this little globe who
refuse to believe in a God, or who, while acknowledging Him, would believe themselves to know
better than He.

THE END

BILLING AND SONS, LTD., PRINTERS, GUILDFORD

WITH ILLUSTRATIONS IN COLOUR
PRICE **6/=** EACH
Large square crown 8vo (6 × 8¼ ins.), cloth, gilt top

By ASCOTT R HOPE
ADVENTURERS IN AMERICA
Containing 12 full-page Illustrations in Colour by HENRY SANDHAM

By Miss CONWAY and Sir MARTIN CONWAY
THE CHILDREN'S BOOK OF ART
Containing 16 full-page Illustrations in Colour from Public and Private Galleries

By ELIZABETH W. GRIERSON
THE CHILDREN'S BOOK OF CELTIC STORIES
12 full-page Illustrations in Colour by ALLAN STEWART

By ELIZABETH W. GRIERSON
THE CHILDREN'S BOOK OF EDINBURGH
Containing 12 full-page Illustrations in Colour from Paintings by ALLAN STEWART

By Mrs. ALFRED SIDGWICK and Mrs. PAYNTER
THE CHILDREN'S BOOK OF GARDENING
Containing 12 full-page Illustrations in Colour by Mrs. CAYLEY-ROBINSON

By G. E. MITTON
THE CHILDREN'S BOOK OF LONDON
Containing 12 full-page Illustrations in Colour by JOHN WILLIAMSON

By G. E. MITTON
THE CHILDREN'S BOOK OF STARS
With a Preface by Sir DAVID GILL, K C.B. Containing 16 full-page Illustrations (11 in Colour) and 8 diagrams in the text

By ELIZABETH W. GRIERSON
CHILDREN'S TALES OF ENGLISH MINSTERS
Containing 12 full-page Illustrations in Colour

By G. E. MITTON
THE BOOK OF THE RAILWAY
Containing 12 full-page Illustrations in Colour by ALLAN STEWART

By ELIZABETH W. GRIERSON
CHILDREN'S TALES FROM SCOTTISH BALLADS
Containing 12 full-page Illustrations in Colour from Paintings by ALLAN STEWART

By S. R CROCKETT
RED CAP ADVENTURES
Being the Second Series of Red Cap Tales Stolen from the Treasure-Chest of the Wizard of the North
16 full-page Illustrations by ALLAN STEWART and others

By S. R CROCKETT
RED CAP TALES
Stolen from the Treasure=Chest of the Wizard of the North
Containing 16 full-page Illustrations in Colour from Drawings by SIMON HARMON VEDDER

By DUDLEY KIDD
THE BULL OF THE KRAAL
A Tale of Black Children
Containing 12 full-page Illustrations in Colour by A. M. GOODALL

By ASCOTT R. HOPE
THE ADVENTURES OF PUNCH
Containing 12 full-page Illustrations in Colour from Drawings by STEPHEN BAGHOT DE LA BERE

A. AND C. BLACK . SOHO SQUARE . LONDON, W.

WITH ILLUSTRATIONS IN COLOUR

(continued)

PRICE **6/=** EACH

Large square crown 8vo. (6 × 8¼ ins), cloth, gilt top

By MIGUEL DE CERVANTES
THE ADVENTURES OF DON QUIXOTE
Translated and Abridged by DOMINICK DALY
Containing 12 full-page Illustrations in Colour from Drawings by STEPHEN BAGHOT DE LA BERE

SWISS FAMILY ROBINSON
Edited by G. E. MITTON
Containing 12 full-page Illustrations in Colour by HARRY ROUNTREE

By JOHN BUNYAN
THE PILGRIM'S PROGRESS
Containing 8 full-page Illustrations in Colour by GERTRUDE DEMAIN HAMMOND, R I

GULLIVER'S TRAVELS
Into Several Remote Nations of the World
By LEMUEL GULLIVER
Containing 16 full-page Illustrations in Colour from Drawings by STEPHEN BAGHOT DE LA BERE

ANIMAL AUTOBIOGRAPHIES

EDITED BY G. E. MITTON

Large square crown 8vo. (6 × 8¼ ins.), cloth, gilt top

THE LIFE STORY OF
A BLACK BEAR
By H. PERRY ROBINSON
With 12 full-page Illustrations in Colour by J. VAN OORT

THE LIFE STORY OF
A FOX
By J. C. TREGARTHEN
With 12 full-page Illustrations in Colour by COUNTESS HELENA GLEICHEN

THE LIFE STORY OF
A CAT
By VIOLET HUNT
With 12 full-page Illustrations in Colour by ADOLPH BIRKENRUTH

THE LIFE STORY OF
A RAT
By G. M. A. HEWETT
With 12 full-page Illustrations in Colour by STEPHEN BAGHOT DE LA BERE

THE LIFE STORY OF
A DOG
By G. E. MITTON
With 12 full-page Illustrations in Colour by JOHN WILLIAMSON

THE LIFE STORY OF
A SQUIRREL
By T. C. BRIDGES
With 12 full-page Illustrations in Colour by ALLAN STEWART

THE LIFE STORY OF
A FOWL
By J W HURST
With 12 full-page Illustrations in Colour by ALLAN STEWART and MAUDE SCRIVENER

The chief feature of this 6/- series, of the 3/6 series on pp. 3-5, and of the 1/6 series on pp. 9 and 10, is the beautiful Illustrations in Colour

A. AND C. BLACK . SOHO SQUARE . LONDON, W

WITH ILLUSTRATIONS IN COLOUR

(continued)

PRICE **6/=** EACH

Large square crown 8vo. (6 × 8¼ ins.), cloth, gilt top

By P G Wodehouse

WILLIAM TELL TOLD AGAIN

Containing 16 full page Illustrations in Colour from Drawings by Philip Dadd

By Harriet Beecher Stowe

UNCLE TOM'S CABIN

Containing 8 full-page Illustrations in Colour from Drawings by Simon Harmon Vedder, in addition to many Illustrations in the text

---

PRICE **6/=**

With 20 full-page Illustrations in Black and White by John Sargent, R.A, and others

GOD'S LANTERN=BEARERS

The Story of the Prophets of Israel for Young People

By Rev. R. C. Gillie

WITH ILLUSTRATIONS IN COLOUR

PRICE **3/6** EACH

Large crown 8vo. (5¼ × 8 ins.), cloth

By John Finnemore

THE STORY OF ROBIN HOOD AND HIS MERRY MEN

Containing 8 full-page Illustrations in Colour by Allan Stewart

By Ascott R. Hope

BEASTS OF BUSINESS

Containing 8 full-page Illustrations in Colour by G Vernon Stokes and Alan Wright

By John Finnemore

THE WOLF PATROL

A Story of Baden-Powell's Boy Scouts

Containing 8 full-page Illustrations in Colour by H. M. Paget

By Lieut.-Col A. F. Mockler-Ferryman

THE GOLDEN GIRDLE

Containing 8 full-page Illustrations in Colour by Allan Stewart

---

A AND C. BLACK . SOHO SQUARE . LONDON, W.

WITH ILLUSTRATIONS IN COLOUR

(continued)

PRICE **3/6** EACH

Large crown 8vo. (5¼ × 8 *ins.*), *cloth*

By John Finnemore

JACK HAYDON'S QUEST

Containing 8 full-page Illustrations in Colour
from Paintings by J. Jellicoe

By Andrew Home

BY A SCHOOLBOY'S HAND

Containing 8 full-page Illustrations in Colour
from Drawings by Strickland Brown

By Hume Nisbet

THE DIVERS

New Edition, containing 8 full-page Illustrations
in Colour from Drawings by the Author

By Andrew Home

EXILED FROM SCHOOL

New Edition, containing 8 full-page Illustrations
in Colour from Drawings by John Williamson

By Ascott R Hope

STORIES

New Edition, containing 8 full-page Illustrations
in Colour from Drawings by Dorothy Furniss

By Andrew Home

FROM FAG TO MONITOR

New Edition, containing 8 full-page Illustrations
in Colour from Drawings by John Williamson

By the Rev. R. C. Gillie

THE KINSFOLK AND FRIENDS OF JESUS

Containing 16 full-page Illustrations in Colour
and Sepia

By the Duchess of Buckingham
and Chandos

WILLY WIND, AND JOCK AND THE CHEESES

Containing 16 full-page Illustrations in Colour
and Black and White, and numerous Pen Drawings in the text by J. S Eland
Square demy 8vo. (6¼ × 9 ins), cloth

By Stanley Waterloo

A TALE OF THE TIME OF THE CAVE MEN

Being the Story of Ab

New Edition, containing 8 full-page Illustrations
in Colour from Drawings by Simon Harmon
Vedder

By Captain Cook

COOK'S VOYAGES AND DISCOVERIES

New Edition, containing 8 full-page Illustrations
in Colour from Drawings by John Williamson

By Daniel Defoe

ROBINSON CRUSOE

Containing 8 full-page Illustrations in Colour
from Pictures by John Williamson

By Mungo Park

TRAVELS IN THE INTERIOR OF AFRICA

To Discover the Source of the Niger

Containing 8 full-page Illustrations in Colour
from Pictures by John Williamson

A. AND C BLACK . SOHO SQUARE . LONDON, W.

THE NEW COLOUR EDITION OF
DEAN FARRAR'S FAMOUS SCHOOL TALES

In large crown 8vo. ·(5¼ × 8 ins.), bound in cloth

Each volume contains 8 full-page Illustrations in Colour and many
in Black and White in the Text

PRICE **3/6** EACH

ERIC ; or, Little by Little

ST. WINIFRED'S ; or, The World of School

JULIAN HOME ; or, A Tale of College Life

The above tales can now be obtained also in three cheaper editions, viz

Large crown 8vo , cloth, with Picture Design and Lettering in Ink, **1/=** each

Large crown 8vo., cloth, with Gold Lettering, Colour Illustration on Cover,
and a Frontispiece, **1/6** each

Large crown 8vo., cloth, Lettered in Gold, with many Black and White
Illustrations, **2/6** each (this was formerly the **3/6** Edition)

By ASCOTT R. HOPE
PEEPS AT THE WORLD
Containing 37 full-page Illustrations in Colour
PRICE **3/6** NET

WITH ILLUSTRATIONS IN BLACK & WHITE

PRICE **3/6** EACH

Large crown 8vo. ´(5¼ × 8 ins.), cloth

By R. S. WARREN BELL
J. O. JONES,
And How He Earned His Living
Containing 12 full-page Illustrations by
GORDON BROWNE

By R S. WARREN BELL
JIM MORTIMER
Containing 16 full-page Illustrations by
GORDON BROWNE

By R. S. WARREN BELL
TALES OF GREYHOUSE
Containing 16 full-page Illustrations by
T. M. R. WHITWELL

By JOHN FINNEMORE
THE STORY OF A SCOUT
Containing 8 full-page Illustrations by
G. E. ROBERTSON

A. AND C BLACK . SOHO SQUARE . LONDON, W.

PRICE **3/6** EACH

Large crown 8vo (5¼ × 8 ins.), cloth

By John Finnemore

TWO BOYS IN WAR-TIME

Containing 6 full-page Illustrations by
Lawson Wood

Edited by T. Ernest Waltham

TANGERINE :

A Child's Letters from Morocco

Containing 78 Illustrations from Photographs
Square demy 8vo. (6¼ × 9 ins.), cloth

By P. G Wodehouse

THE GOLD BAT

Containing 8 full-page Illustrations from
Drawings by T. M R. Whitwell

By P. G. Wodehouse

THE POTHUNTERS

Containing 10 full-page Illustrations by
R. Noel Pocock

By P. G. Wodehouse

A PREFECT'S UNCLE

Containing 8 full-page Illustrations by
R. Noel Pocock

By Sir Clements Markham, K.C.B.

THE PALADINS OF EDWIN THE GREAT

Containing 10 full-page Illustrations by
Ralph Peacock

By the Rev. R. C. Gillie

THE STORY OF STORIES

A Life of Christ for the Young

Cheap reissue
Containing 32 full-page Illustrations

By P. G. Wodehouse

TALES OF ST. AUSTIN'S

Containing 12 full-page Illustrations by
T M R. Whitwell, R. Noel Pocock, and
E. F Skinner

By P G. Wodehouse

THE HEAD OF KAY'S

Containing 8 full-page Illustrations by
T. M. R. Whitwell

By P. G Wodehouse

THE WHITE FEATHER

Containing 12 full-page Illustrations by
W. Townend

By P. G. Wodehouse

MIKE :

A Public School Story

Containing 12 full-page Illustrations by
T. M R. Whitwell

A. AND C. BLACK . SOHO SQUARE . LONDON, W.

BOOKS BY ASCOTT R. HOPE

HERO AND HEROINE

Containing 9 Illustrations by A. Hitchcock

Crown 8vo , cloth

PRICE **5/=**

CAP AND GOWN COMEDY

HALF= TEXT HISTORY

PRICE **3/6** EACH

Without Illustrations

PRICE **3/6** EACH

BLACK AND BLUE

Illustrated, crown 8vo (5 × 8 ins.), cloth

ALL ASTRAY

Containing 11 Illustrations by W. H. C Groome

Crown 8vo , cloth

Crown 8vo., cloth, without illustrations

PRICE **5/=** EACH

AN ALBUM OF ADVENTURES

THE SCHOOLBOY ABROAD

READY= MADE ROMANCE

DRAMAS IN DUODECIMO

TWO BEAUTIFUL BOOKS

ALTHOUGH THESE TWO BEAUTIFULLY ILLUSTRATED BOOKS ARE PRIMARILY INTENDED FOR THE SCHOOLROOM, THEY ARE YET WRITTEN AND ILLUSTRATED IN SUCH A MANNER AS TO RENDER THEM EMINENTLY SUITABLE FOR GIFT-BOOKS, AND THE GROWN-UPS WHO TURN OVER THE LEAVES WILL SURELY FEEL REGRET THAT THE GEOGRAPHY OF THEIR YOUTHFUL DAYS WAS NOT PRESENTED IN SO DELIGHTFUL A FORM

PRICE **1/6**

THE CHILD'S WORLD IN PICTURES

By C. von Wyss

Containing 62 Illustrations, 32 of which are in Colour

Crown 4to , Picture Boards

PRICE **2/6**

THE CHILD'S GEOGRAPHY OF ENGLAND AND WALES

By L. W Lyde

Containing 32 full-page Illustrations in Colour

Square crown 8vo , cloth

A. AND C. BLACK . SOHO SQUARE . LONDON, W

THE WAVERLEY NOVELS

(VICTORIA EDITION)

BY SIR WALTER SCOTT

Price **1/6** per Vol.

In 25 vols., crown 8vo. (5 × 8 ins.), bound in cloth, and Illustrated with Frontispieces

THE Authentic Editions of Scott are published solely by A. and C. BLACK, who purchased along with the copyright the interleaved set of the Waverley Novels, in which Sir Walter Scott noted corrections and improvements almost to the day of his death. The Victoria Edition and the editions given on page 12 have been collated word for word with this set, and many inaccuracies, some of them ludicrous, corrected.

LIST OF THE NOVELS

Waverley
Guy Mannering
The Antiquary
Rob Roy
Old Mortality
Montrose, and Black Dwarf
The Heart of Midlothian
The Bride of Lammermoor
Ivanhoe
The Monastery
The Abbot
Kenilworth
The Pirate

The Fortunes of Nigel
Peveril of the Peak
Quentin Durward
St. Ronan's Well
Redgauntlet
The Betrothed, etc.
The Talisman
Woodstock
The Fair Maid of Perth
Anne of Geierstein
Count Robert of Paris
The Surgeon's Daughter, etc.

For other Editions of Scott see page 12

A. AND C BLACK . SOHO SQUARE . LONDON, W.

TRAVEL BOOKS FOR YOUNG PEOPLE WITH ILLUSTRATIONS IN COLOUR

PRICE 1/6 NET

EACH CONTAINING 12 FULL-PAGE ILLUSTRATIONS IN COLOUR

Post free, 1/9.   *Large crown 8vo* (5¼ × 7¾ *ins.*), *cloth.*   **Post free, 1/9**

PEEPS AT MANY LANDS

THIS is a series of "Little Colour Books" for Little Readers, and is meant to give children a glimpse at the scenes and customs of their own and other lands. Each book is written in a simple and interesting style, and nowhere smacks of the geographical text-book The little volumes are not designed as lesson-books, though much may be learned from them , but the aim is that the child shall gain this instruction through the sheer pleasure of reading. A strong feature is made of the work and play of children in the land under description, and the general ways of life among the people form another special point.

The volumes are handsomely bound and beautifully illustrated in colour.

AUSTRALIA. By Frank Fox.   *[In preparation*

BELGIUM. By George W. T Omond. With 12 Illustrations by Amédée Forestier.

BURMA By R. Talbot Kelly, R.B.A , F.R.G.S., Commander of the Medjidieh With 12 Illustrations by Mr Kelly.

CANADA. By J. T. Bealby. With 12 Illustrations by T. Mower Martin and others.

CEYLON. By Alfred Clark.   *[In preparation.*

CHINA. By Lena E Johnston With 12 Illustrations by Norman H. Hardy.

CORSICA. By Ernest Young, B Sc. With 12 Illustrations by E. A Norbury, R.C A.

DENMARK. By M. Pearson Thomson.   *[In preparation.*

EGYPT. By R Talbot Kelly, R B A , F.R G S., Commander of the Medjidieh. With 12 Illustrations by Mr. Kelly.

ENGLAND· By John Finnemore. With 12 Illustrations by various artists.

FINLAND. By M. Pearson Thomson. With 12 Illustrations by Allan Stewart and others.

A AND C. BLACK . SOHO SQUARE . LONDON, W.

PEEPS AT MANY LANDS

(*continued*)

FRANCE. By .John Finnemore   With 12 Illustrations by E. Crescioli and others

GERMANY. By Mrs Alfred Sidgwick. With 12 Illustrations by L Burleigh Bruhl and others.

GREECE. By E. A. Browne. With 12 Illustrations by E. H. Fitchew and John Fulleylove.

HOLLAND. By Beatrix Jungman. With 12 Illustrations by Nico Jungman.

HOLY LAND. By John Finnemore   With 12 Illustrations by John Fulleylove.

ICELAND By Mrs. Disney Leith   With 12 Illustrations by M. A C. Wemyss and the Author.

INDIA. By John Finnemore. With 12 Illustrations by Mortimer Menpes

IRELAND. By Katharine Tynan   With 12 Illustrations by Francis S Walker

ITALY. By John Finnemore. With 12 Illustrations by Alberto Pisa and others.

JAMAICA. By John Henderson   With 12 Illustrations by A. S Forrest

JAPAN. By John Finnemore   With 8 Illustrations by Ella du Cane.

MOROCCO. By John Finnemore. With 12 Illustrations by A. S. Forrest

NEW ZEALAND. By P A. Vaile. With 12 Illustrations by F and W. Wright.

NORWAY. By Lieut.-Col. A. F. Mockler-Ferryman, F R.G.S., F.Z.S. With 12 Illustrations by A. Heaton Cooper and Nico Jungman.

PORTUGAL. By A. M. Goodall. With 12 Illustrations by the Author.

RUSSIA. By L. E. Walter.   [*In preparation.*

SCOTLAND. By Elizabeth Grierson. With 12 Illustrations by William Smith, junr., and others

SIAM By Ernest Young. With 12 Illustrations by Edward A. Norbury.

SOUTH AFRICA By Dudley Kidd   With 12 Illustrations by Agnes M. Goodall.

SOUTH SEAS By J H. M. Abbott. With 12 Illustrations by Norman Hardy.

SPAIN. By Edith A. Browne.   [*In preparation.*

SWITZERLAND. By John Finnemore. With 12 Illustrations by J. Hardwicke Lewis and A. D. McCormick

TURKEY. By John Finnemore.   [*In preparation.*

WALES. By Jeanette Marks.   [*In preparation.*

THE WORLD. By Ascott R. Hope. With 37 Illustrations in colour by various artists and a sketch map. Price **3s. 6d.** net.

A AND C. BLACK . SOHO SQUARE   LONDON, W.